PENGUIN BOOKS

IN THE BEGINNING

Dr John Gribbin trained as an astrophysicist at the University of Cambridge before becoming a full-time science writer. He has worked for the science journal *Nature* and the magazine *New Scientist* (for which he is now physics consultant) and has contributed articles on science topics to *The Times*, the *Guardian* and the *Independent*. Gribbin has received awards for his writing in both Britain and the United States and is currently a visiting Fellow in astronomy at the University of Sussex. His books include *In Search of Schrödinger's Cat*, *In Search of the Big Bang* and *The Matter Myth* with Paul Davies, and the bestselling books *Stephen Hawking: A Life in Science* and *Einstein: A Life in Science* written with Michael White. He is also the author of several science fiction works, most recently *Innervisions*.

Married, with two sons, John Gribbin lives in East Sussex.

JOHN GRIBBIN

IN THE BEGINNING

THE BIRTH OF THE LIVING UNIVERSE

PENGUIN BOOKS

PENGUIN BOOKS

Published by the Penguin Group
Penguin Books Ltd, 27 Wrights Lane, London W8 5TZ, England
Penguin Books USA Inc., 375 Hudson Street, New York, New York 10014, USA
Penguin Books Australia Ltd, Ringwood, Victoria, Australia
Penguin Books Canada Ltd, 10 Alcorn Avenue, Toronto, Ontario, Canada M4V 3B2
Penguin Books (NZ) Ltd, 182–190 Wairau Road, Auckland 10, New Zealand

Penguin Books Ltd, Registered Offices: Harmondsworth, Middlesex, England

First published by Viking 1993
Published in Penguin Books 1994
1 3 5 7 9 10 8 6 4 2

Printed in England by Clays Ltd, St Ives plc

Contents

Preface

Two of my main scientific interests have long been cosmology, the study of the Universe at large, and evolution, the theory which explains our presence here on Earth. During most of the twenty years or so that I have been writing books, these seemed widely disparate interests, linked only by the fact that both are subject to probing by human inquisitiveness, in the quest to find out as much as possible about both ourselves and the Universe we inhabit. But while working on my two most recent books, I became convinced that there is much more to the story of life and the Universe than meets the eye. In *The Matter Myth*, Paul Davies and I discussed the nature of complexity, and how simple physical laws, operating on seemingly simple physical systems, can give rise to behaviour that is more complex than the sum of its constituent parts. In *In Search of the Edge of Time* (*Unveiling the Edge of Time* in the USA), I set out to tell the story of black holes in the Universe, and ended up being led by recent developments in cosmology to the idea that our Universe is just one among many. And then, in the spring of 1992, came news that the COBE satellite had discovered fresh evidence supporting the Big Bang theory, and the notion that the Universe was born at a definite moment in time.

The discovery acted as a catalyst, helping me to put my fumbling thoughts about life and the Universe in order. If the Universe itself was born, and will one day die, and if it is just one Universe among many, and if things are in general more complex than simple physical laws suggest, then the theory of evolution might be much more relevant to the story of cosmology than I had previously suspected. The Universe might, indeed, be alive – literally, not metaphorically – in its own right.

This book, which can be seen as a sequel to both *Matter Myth* and *Edge of Time*, is my attempt to explain the connection between evolution and cosmology, and to point the way to a new understanding of the Universe, an understanding being developed after COBE but which tells us what went on *before* the Big Bang. The ideas I describe here represent the biggest shift in our understanding of the Universe, and of our own place in the Universe, since the Big Bang theory itself was first discussed more than half a century ago. It explains how and why the Universe itself came into existence, and how and why we came into existence. You cannot really ask for much more of any scientific idea; and it all begins, as all good stories should, in the beginning – with COBE's evidence for the birth of the living Universe.

John Gribbin
December 1992

PROLOGUE

The End of the Beginning

Astronomers breathed a huge collective sigh of relief in the spring of 1992, when the most important prediction they had ever made was proved correct at the eleventh hour. The dramatic discovery of ripples in the structure of the Universe dating back almost 15 thousand million years has set the seal on twentieth-century science's greatest achievement – the Big Bang theory which explains the origin of the Universe and everything in it, including ourselves. Slotting in this missing piece of the cosmic jigsaw puzzle confirms that the Universe really was born out of a tiny, hot fireball all that time ago, and has been expanding ever since. But this is not the end of the story of cosmic creation and the evolution of the Universe. It is scarcely even the end of the beginning of that story, for this discovery also points the way to a new understanding, not just of the origin but (literally) of the evolution of the cosmos.

The story really dates back to the 1950s, when there were two rival theories to explain the nature of the Universe. Astronomers knew that the Universe contains many millions of galaxies, each one, like our own Milky Way, made up of billions of stars. (In this book, 'billion' means a thousand million.) And they knew that these galaxies are moving away from one another, as the empty space between the galaxies expands. The simplest interpretation of this cosmic expansion is that long ago the Universe must have been much smaller, because there was less space between the galaxies. Take this notion to its limit, and you have the idea of the Big Bang – that everything in the Universe emerged from a point, known as a singularity, some 15 billion years ago.

But the rival theory, known as the Steady State hypothesis, said that the Universe might have been expanding forever, without changing its overall appearance. The chief proponent of this hypothesis was the British astronomer Fred Hoyle (now Sir Fred). According to this idea, as the space between the galaxies stretched, new galaxies were continuously being created, out of nothing at all, to fill in the gaps. Of course, this would require the continual creation of new matter, in the form of hydrogen atoms, in the empty space between galaxies in order to provide the building material for the stars of new galaxies. Many scientists were horrified at the prospect; but Hoyle and his colleagues argued that this rather gentle form of continuous creation was no more abhorrent, in principle, than the notion that *all* of the matter in the Universe had been created in one instant, the Big Bang. Indeed, Hoyle himself coined the expression 'Big Bang' as a term of derision for a theory that he once described as being about as elegant as a party girl jumping out of a cake.

In the 1950s, which theory you preferred was largely a matter of personal prejudice. Philosophically, is it more difficult to accept that matter is created continually in the Universe in small dribbles, or that all of the matter in all the stars and galaxies was created at a singular fixed moment in time?

The conflict between the two rival theories was resolved in the early 1960s by a dramatic and unexpected discovery. Two American radio astronomers, Arno Penzias and Robert Wilson, working with a radio antenna owned by the Bell Laboratories, discovered a weak hiss of radio noise coming from all directions in space. This radio noise, now dubbed the 'cosmic background radiation', was quickly explained as the remnant of the fireball in which the Universe was born, the echo of the Big Bang itself.

As the Universe has expanded for 15 billion years, the hot radiation in the original fireball has expanded with it, and cooled as a result. (This is exactly equivalent to the opposite *heating* effect that warms the air in a bicycle pump when you pump up the tyres of a bicycle; when things are compressed they get hot, when they expand they cool.) When astronomers measured the temperature of this radiation, they found that it was just under −270°C. This is exactly the temperature the radiation ought to have if the universal expansion has proceeded since the Big Bang in line with Albert Einstein's general theory of relativity.

But there was a snag. In the 1970s and 1980s, astronomers became concerned that the cosmic background radiation is *too* smooth and uniform. The radiation is left over from the last phase of the Big Bang fireball, about 300,000 years after the Big Bang singularity itself. If the radiation coming from all parts of the sky is perfectly smooth, that means that the Universe was perfectly smooth just 300,000 years after the Big Bang. In a *perfectly* smooth Universe, as will become clear as my story unfolds, there would be no stars or planets, and therefore no people. And yet, we know that the Universe is full of galaxies, galaxies are full of stars, and stars are orbited by planets. We live on one of those planets. If the Universe were perfectly smooth, we would not exist.

Late in 1989, the COBE satellite (the acronym stands for Cosmic Background Explorer) was launched to find out just how smooth the background radiation really is. The Big Bang calculations said that the ripples in the background radiation needed to explain the existence of galaxies – and of ourselves – must be there, but that only receivers in space, above the interference of the Earth itself and artificial radio noise, would be able to detect them.

At first the COBE measurements seemed to show a Universe too smooth to allow galaxies to exist. But the theorists maintained that the ripples required to confirm the Big Bang theory were so small that it would take two years of painstaking observation by COBE for them to be picked out. They would correspond to temperature fluctuations in the radiation from different parts of the sky of only about 30 millionths of a degree.

The months that followed were a nail-biting time for those theorists. If COBE still had not found any fluctuations by the middle of 1992, the whole Big Bang theory might have had to be rejected. But in April 1992, right on cue, came the news from

NASA that ripples in the cosmic background radiation of exactly the right size had indeed been detected. The discovery, announced at a meeting of the American Physical Society in Washington, DC, was described by the NASA team as 'evidence for the birth of the Universe'.

The ripples in the background radiation confirm that 300,000 years after the Big Bang singularity itself there were already wispy clouds of matter stretching across vast distances, upwards of 500 million light years across. The size of these clouds also confirms that more than 90 per cent of the Universe today is in the form of so-called 'dark matter', not the visible bright stars and galaxies. As those clouds collapsed in upon themselves, pulled together by their own gravity, they would have broken up and formed clusters of galaxies, with the galaxies themselves breaking up into stars like those of the Milky Way, all embedded in a sea of dark stuff. The tiny temperature fluctuations measured by COBE are coming from the seeds of our own existence.

But this is not the end of the Big Bang story. Emboldened by this evidence that their theories were right all along, astronomers are already speculating on what happened before the Big Bang. If the Universe was born, what was it born out of – what was its parentage? And how will it die? Did the Universe really appear as a singularity bursting out from nothing at all? Or was there a previous cycle in which that singularity was created by the collapse of an earlier phase of the Universe into a black hole? Will the Universe expand forever, or will it one day recollapse, forming a new singularity out of which a new Universe might burst? With the Big Bang theory itself on a secure foundation, it is questions like these that now provide scope for cosmological speculation, and which will be making headline news in the years and decades to come.

These questions, and the present-day attempts to answer them, form the theme of this book. As we shall see, terms like 'evolution', 'life' and 'death', and perhaps even 'parent' and 'baby', may have far more than a metaphorical application to the Universe in which we live, and to some of the structures within that Universe.

The bursting-out of the Universe from the Big Bang is *precisely* equivalent, within the framework of the general theory of relativity, to the time-reversed 'mirror-image' of the collapse of a massive object into a black hole. This was proved by Roger Penrose and Stephen Hawking, at the end of the 1960s. So could the collapse of a previous cycle of the Universe towards a singularity have been, literally, the collapse of a black hole?

Several researchers have investigated this possibility and its implications. Their extraordinary discovery is that the collapse of a black hole – *any* black hole – can indeed lead to a 'bounce' which creates a new universe. But there are two exciting new implications of this work.

First, you do not need to collapse the entire Universe in order to make a new one. Even if a black hole formed from a star only three or four times bigger than our Sun, because of the way the laws of physics are changed at the singularity it will bounce to form a full-scale Universe in its own right. But *where* will this Universe be? It doesn't come bouncing out from the black hole into our world, but expands away in a new set of dimensions of its own!

Secondly, every time a universe is created in a big bang bounce in this way, the laws of physics that it is born with are slightly different from those of its parent universe – they mutate. The force of gravity, for example, may be a little stronger (or a little weaker) in the 'baby' universe than in the 'parent'.

Together, these discoveries have led to the idea that our Universe is just one among a multitude of universes, and that in some sense the many universes are competing with one another for the right to exist. Is it possible, then, that this competition between universes might follow the rules of competition between species here on Earth, in the Darwinian sense? Is our Universe alive, and has it evolved by natural selection?

These are the questions I shall address in this book. Answering those questions requires a synthesis of the ideas of biological evolution with those of cosmology, traditionally the domain of the physicist. It is the synthesis, rather than any individual strand, that makes this work so exciting, and points the way to a new cosmology of the twenty-first century. Before we can consider the synthesis itself, we shall have to decide exactly what we mean by 'life', and what we mean by 'the Universe'. And before I come to those intriguing questions, it is only right and proper to put the COBE discoveries in perspective, and to commence my story of the living Universe, as I promised in the Preface, where all good stories should start: in the beginning – which means, in this case, the very beginning of the entire Universe as we know it.

PART ONE

The Birth of the Universe

CHAPTER ONE

Our Changing Universe

We live in a dynamic, changing Universe. It was born, and it will die. Everything within the Universe, including the Sun and the stars, has its own life cycle. Nothing is eternal. You might think that this is the most natural and obvious way for things to be. After all, we are living creatures, used to the cycle of birth, life and death which we see all around us on the planet we call Earth. And yet the realization that the Universe itself is not eternal, but had a definite beginning at one moment in time, dates back less than seventy years – less than the Biblical 'three score years and ten' of a human lifetime. The realization was so startling, and so profound, that it has taken that full human lifetime for the implications to sink in, and for scientists to begin to address properly the questions raised by the discovery – questions concerning the life and death not just of our Universe, but possibly of other universes as well.

At the beginning of the 1920s, astronomers were only just beginning to comprehend that the stars we can see in the sky represent only a tiny fraction of the Universe. With the aid of new telescopes, they discovered that the entire Milky Way system, which is made up of perhaps a hundred billion stars, each more or less like our Sun, is just one island in space, a galaxy. Beyond the Milky Way there are millions of other galaxies, scattered through the blackness of space like coral islands scattered across the Pacific Ocean. Each of those galaxies, appearing as no more than a fuzzy blob of light even with the aid of the best telescopes on Earth, may contain as many individual stars as our own local Galaxy, the Milky Way.

The first breathtaking leap out into the Universe from the Milky Way came in the mid-1920s, when studies of variable stars showed that a fuzzy blob then known as the Andromeda Nebula is actually a galaxy, now known to be some two million light years away from us. For comparison, the whole of the Milky Way system is a flattened disk about 100,000 light years across, and the Sun with its family of planets lies about 30,000 light years out from the centre, very much in the backwoods of our home Galaxy. (A light year is the distance travelled by light, at the headlong speed of 300 million metres *per second*, in one year.) Light from the Andromeda Galaxy takes more than two million years to cross the intervening space to our telescopes. Yet this breathtaking leap out into space was still only a modest first step on the cosmic scale of things, and improving observations soon showed that many galaxies are much farther away than this, at distances of tens or hundreds of millions of light years. The whole Milky Way system in which we live suddenly shrank, in the astronomical imagination, to a tiny mote floating in a vast sea of emptiness. Over the years following this discovery, Edwin Hubble, in California, and other astronomers measured the distances to other galaxies out to about a thousand million light years – a volume of space encompassing some hundred million galaxies, although only a handful have ever been studied in detail.

The key to measuring cosmic distances across millions of light years was the discovery that some variable stars change in brightness in a regular way, and that the length of the cycle of these changes depends on the average brightness of the star – they obey what is called a 'period–luminosity law'. The most important of these distance indicators are called Cepheid variables. An individual Cepheid will first brighten, then dim, then brighten again. The period of this variation may be anything from two to forty days, depending on the individual star, but it is always the same for each individual star, and the exact length of the period – say, 22 days – tells us exactly how bright the star really is. The apparent brightness (or dimness) of the star as seen from Earth then reveals its distance, since the apparent brightness is just the actual brightness divided by the square of the distance (so, for example, if a star is twice as far away, it is only a quarter as bright).

By 1929, there were enough distance measurements to other galaxies for Hubble to notice a curious phenomenon. Apart from our very nearest neighbours, such as the Andromeda Galaxy (at a distance of 'only' 2 million light years), all the galaxies seemed to be moving away from us. What was more, the speed with which each galaxy was moving away (its recession velocity) was proportional to the distance from us (so a galaxy twice as far away was receding from us twice as fast). The farther away a galaxy was, the faster it was moving away.

Hubble knew the distances to some galaxies from the Cepheid rule, and he knew how fast they were moving from measurements of their spectra. The light from any object can be split up into the colours of the rainbow – a spectrum – using a prism. You can do the trick yourself using a triangular wedge of glass or clear plastic to make rainbow patterns with sunlight. When such a spectrum is studied carefully using a combination of a microscope and a prism (a spectroscope), the bright rainbow colours are seen to be crossed by narrow lines, some dark and some bright. These lines are characteristic of the atoms that are radiating the light, and they are as unique as fingerprints, with one set of lines for each kind of atom (each element). The lines occur at very precise wavelengths which can be measured in light from objects studied in laboratories here on Earth; the same element (hydrogen, or oxygen, or carbon, or whatever) always produces the same pattern of lines at the same wavelengths. Light from distant galaxies shows the same kinds of fingerprint pattern of spectral lines as light from the Sun and the stars of the Milky Way. But the whole pattern of lines is shifted bodily towards the red end of the spectrum – a phenomenon known as the 'redshift'.

The colours of the spectrum of visible light are, in order, red, orange, yellow, green, blue, indigo and violet. Of these, red light has the longest wavelength and violet the shortest, although the spectrum actually extends in both directions beyond the range that our eyes can see. In one direction, longer-wavelength radiation includes infrared, microwaves and radio waves; in the other, shorter-wavelength radiation includes ultraviolet, X-rays and gamma-rays. The redshift in the light from distant galaxies means that the light has a longer wavelength than it 'ought' to have; it has been stretched on its journey across space to us. How could this happen?

The simplest explanation is that those distant galaxies are moving away from us. The waves in the light from an object that is receding in this way are indeed stretched and redshifted, and the amount of redshift reveals the speed at which the object is moving. In a similar way, an object moving towards us emits light which is blueshifted by its motion, with the waves squashed together. The whole process is rather like the way in which the pitch of the note from a siren on an ambulance or police car seems to be deeper if the vehicle is rushing away from you, but higher if the vehicle is approaching you. This is known as the Doppler effect, and it occurs because sound waves are squeezed and stretched in just the way I have described for light; deeper notes have longer wavelengths than higher notes.

You might expect that if all the galaxies are moving at random through the Universe, about half of them would be coming towards us and half moving away, so that astronomers would detect roughly equal numbers of redshifts and blueshifts. But Hubble discovered that, outside our immediate cosmic neighbourhood, there are no blueshifts to be seen in the spectra of galaxies. There are only redshifts, and the redshift of a galaxy is proportional to its distance, a property which is now known as Hubble's law.

Hubble's law, discovered at the end of the 1920s, did two things. First, once the law had been calibrated on relatively nearby galaxies, using the Cepheid method, it meant that from then on astronomers had only to measure the redshift to a galaxy in order to determine its distance – even if that galaxy was so far away that any Cepheids in it were too faint to be picked out individually. Distances across the Universe could now confidently be measured in billions of light years. But the second implication of Hubble's law was even more dramatic. It said that the entire Universe was expanding – that every galaxy is moving apart from every other galaxy. As time passes, galaxies get farther apart and the space between them widens. Which means, of course, that if we look *back* in time, in our imagination, galaxies used to be closer together, with less space between them. Taking this discovery to its logical conclusion, there must have been a time when all the galaxies were on top of one another, with *no* space between them. It was this interpretation of Hubble's law that led to the idea of the Big Bang – the idea that the Universe was born in a superdense fireball at a definite moment in time.

But perhaps the most extraordinary thing about Hubble's law is that both the law and the fact that the Universe is expanding had been predicted, a dozen years before, by Albert Einstein's general theory of relativity. The only snag was that Einstein himself had not trusted the equations. Far from publicizing the prediction, he had actually dropped in an extra term, a 'fudge factor', to cancel out the expansion that his equations predicted. He later described this as the 'biggest blunder' of his scientific career; but in 1917, the notion that the Universe was unchanging and eternal was so deeply ingrained that even Einstein preferred to tamper with his theory rather than accept what it told him. Once Hubble's work showed that the Universe really is expanding, however, the un-fudged version of Einstein's equations soon provided cosmologists with the tool they needed to explain what was going on.

Einstein's general theory of relativity explains and describes the behaviour of space, time and matter. In other words, it describes everything in the Universe. Einstein had already explained the nature of space and time, in 1905, with his special theory of relativity. This is the theory that tells us, among other things, that nothing can ever travel faster than light, and that the speed of light (*c*) is an absolute constant. Curious though it may seem, this means that if I am travelling north at half the speed of light and you are travelling south at half the speed of light, if you flash a signal to me with your torch *both* of us will measure the speed of that light signal as c. You do not see the light beam moving forward at only $0.5c$, even though you are moving after it at half the speed of light, and I do not see the light beam approaching me at $1.5c$, even though I am running head on into it at half the speed of light. Einstein also discovered that mass and energy are interchangeable ($E = mc^2$). It is important to appreciate that all of this has been tested and confirmed many times by experiment; it may not be common sense, but Einstein's description of the Universe has come through every test with flying colours, and is without doubt a good description of what is going on out there.

The key feature of the special theory is that it combines space and time into one package, known as spacetime, described by one set of equations. In the world of relativity, time is a dimension, like the familiar three dimensions of space (up–down, left–right, forward–backward). The three dimensions of space are all at right angles to one another; we cannot visualize adding in a fourth dimension at right angles to all three of them, because our brains do not work that way. But the equations tell us that spacetime really is like that, with the added complication that the time dimension somehow involves *negative* distances – there is a minus sign in front of the time parameter in the equations.

Happily, we do not have to worry about the mathematical details, because the relativists have come up with a simple, easily pictured 'model' of what is going on. Instead of trying to think in four dimensions, we can think in two. Imagine spacetime as being like a flat, stretched rubber sheet – like the skin of a drum, or a trampoline. We can ignore two of the space dimensions, since they are all the same as one another, mathematically speaking, and imagine that one direction across the stretched sheet represents movement through space, while the direction at right angles to this represents movement through time. Roll a marble across the sheet, and you have a picture of the trajectory of an object through spacetime.

This is where the general theory of relativity comes in. The special theory deals only with flat spacetime (flat trampoline sheets). Einstein spent ten years, off and on, trying to include the effect of gravity in his theory (which means adding in the influence of matter), and he announced the successful completion of this task in 1915. Again, leaving aside the equations and thinking in terms of the geometry of spacetime, we can understand his improved theory by imagining what happens to our stretched trampoline if we put a heavy weight on it. The stretched sheet bends, forming a dip with the weight at the centre. Now, if we roll a marble across the sheet it will follow a curved trajectory, into and out of the dip. This is what matter does to four-dimensional spacetime in the real Universe. Spacetime is bent and distorted by the presence of heavy objects (like the Sun), and anything passing near the heavy object will follow a curved trajectory through the distorted region of spacetime.

Again, all this has been tested and measured to many decimal places. As early as 1919, astronomers were able to measure how light itself is bent as it passes near the Sun, using photographs of the stars behind the Sun obtained during a total solar eclipse, when the dazzling light from the Sun itself was blotted out by the Moon. The amount of light-bending they measured had been predicted, exactly, by Einstein's theory. There is no doubt at all that the general theory of relativity is a good description of how the Universe works. The way to remember what is going on is that matter tells spacetime how to curve, and the curve in spacetime tells matter how to move. What we think of as the force of gravity, making objects fall to the ground or holding planets in orbit around the Sun, is actually the curvature of spacetime, deflecting moving objects from the straight-line path they would follow if spacetime were flat.

All this is intriguing enough. But the big surprise came in 1917, when Einstein used his equations to produce a mathematical description of the Universe itself. How would spacetime bend, and how would matter move, if the Universe were uniformly filled with matter? This was the point at which Einstein refused to believe what his own equations were telling him. For they said that under those circumstances space itself would not just bend, but stretch – the Universe would expand, not because the lumps of matter in it were moving away from one another *through* space, but because the space itself was stretching, as if the sides of the trampoline in the two-dimensional 'model' of the Universe were being moved farther apart, steadily stretching the rubber of the trampoline's surface. In fact, the equations also said that the Universe might be shrinking, with space contracting steadily; but the one thing the equations did not allow for was the possibility of a static Universe, unchanging and eternal.

Einstein's decision to add an extra term to his equations to hold the Universe still is understandable, from the perspective of 1917. But as soon as Hubble had discovered the expansion of the real Universe in which we live, that fudge factor was redundant, and cosmologists were able to take over Einstein's equations as a description of the real Universe.

The key feature of this description has already been mentioned – galaxies are *not* moving away from one another by travelling through space. Instead, they are being carried apart as the space between the galaxies stretches. Individual galaxies do indeed have a certain amount of random motion through space, and this is why our nearest large neighbour, the Andromeda Galaxy, is actually moving towards us and shows a blueshift. But once we look out beyond our immediate neighbourhood, these local, random velocities are much smaller than the recession velocities caused by the expansion of space itself. The process produces exactly the same redshift law, with redshift proportional to distance, but this is now understood not as a genuine Doppler effect but as being literally due to the light waves from distant objects being stretched to longer wavelengths on their journey to us.

There is another very important feature of the Einstein–Hubble Universe to take on board. When we look out into space, we see distant galaxies receding uniformly in all directions, as if they are fleeing from our own Milky Way Galaxy. But this does *not* mean that we live at the centre of the Universe. Hubble's law, with redshift proportional to distance, is the only redshift–distance law that would give exactly the same picture of the expanding Universe from whichever galaxy you happened to be sitting on. One way to picture this is by imagining a series of inkspots marked on the surface of a balloon. If you blow the balloon up to twice its original size, the distance between every pair of spots doubles. So the farther apart two spots were to start with, the more they separate, exactly mimicking the redshift rule found by Hubble. Two spots that used to be 2 cm apart are now 4 cm apart, while two spots that used to be 4 cm apart are now 8 cm apart. But you cannot point to any one spot and say that it is at the centre of the pattern. Whichever spot you measure from, you will always find that the other spots are now farther away than before.

Another analogy is with the raisins in a loaf of raisin bread, being carried apart from one another as the dough rises. This analogy is not perfect, because a loaf of bread does have a centre, and an edge. Einstein's equations tell us that the actual Universe may have no edge at all, because space is gently curved around upon itself to make the equivalent, in four dimensions, of the surface of a sphere. Like the surface of the Earth, or the surface of the spotted balloon, the space we live in may be 'closed', with no edge. If you set off on a journey here on Earth and keep going in the same direction, you will eventually travel round the Earth and back to where you started. In the same way, if you set out on a journey through space and keep going in the same direction, you will eventually travel around the Universe and return to your starting-point. There is no centre to the Universe, just as there is no centre to the surface of the Earth or to the skin of a soap bubble.

During the 1930s and 1940s, cosmologists began to try to come to terms with the implications of all this. The most revolutionary aspect of the new cosmology was the implication that the Universe must have been born at a definite moment in time. Winding back the expansion of the Universe, you could imagine space shrinking away and galaxies piling up on top of one another. Taken at face value, Einstein's equations said that everything in the Universe (space, time and matter) must have emerged from a single point, a singularity.

At first, nobody took the idea of the singularity too seriously. But they did try to imagine, and describe mathematically, a time when all of the matter in the Universe would have been squeezed into one lump, a kind of 'primeval atom' or 'cosmic egg', which then burst forth and became the Universe as we know it today, stars and galaxies forming out of the debris and galaxies being carried apart as the space between them expanded. (Remember, though, that this primeval explosion was *not* the result of a superdense lump of matter sitting in the midst of empty space and then exploding outward into space, like the explosion of a bomb; empty space itself was also squeezed down into the primeval lump, and expanded outward from the birth of the Universe.) The first big question the cosmologists were faced with in the 1930s was, if the Universe was born at a definite moment in time, *when* was it born?

It is very easy to work out when all the galaxies were on top of each other, from Hubble's law. Provided the distance measurements based on Cepheid variables are right, you just have to divide the distance to a galaxy by its recession velocity (astronomers still use the term 'recession velocity', even though they know that the redshift is actually caused by space itself expanding) to find out how long it has taken for that galaxy to get so far from us. This is exactly the same as the way in which you calculate that a car that has been driven at 100 kph (60 mph) and has covered a distance of 300 km (180 miles) must have been on the move for three hours. Because of the simple nature of Hubble's law, with redshift (recession velocity) proportional to distance, you get the same answer for every galaxy.

When Hubble himself carried out the calculation in the 1930s, the 'answer' he came up with implied that the Universe was only about two billion years old. But in the 1940s astronomers realized that there are actually two different kinds of Cepheid variable, each with its own period–luminosity law, and that Hubble had confused the two in his calculations. Taking advantage of the wartime blackout to study very faint stars in the Andromeda Galaxy through the 100-inch telescope on Mount Wilson, near Los Angeles, Walter Baade refined the distance scale and more than doubled the distance estimate to every galaxy. Over the past half-century, estimates of the distance scale and age of the Universe have been constantly refined, and the best estimate today is that the time that has elapsed since the Big Bang is somewhere between 13 billion and 20 billion years.

In round terms, the Universe is some 15 billion years old. For comparison, our Solar System, including the Earth and the Sun itself, is a little less than 5 billion years old, and has been around for roughly one-third of the life of the Universe so far. The most distant objects known, quasars with very high redshifts, are so far from us that the light by which we see them set out on its journey when the Universe was only 10 per cent of its present age; the light has spent more than ten billion years on its journey (twice as long as the age of our Solar System), and in that sense the most distant objects we can detect with modern optical telescopes are more than 10 billion light years away. But astronomers can detect radiation from even farther away, and even further back in time. This radiation is not in the form of light, but of radio waves. It started out as light, and high-energy radiation such as X-rays and gamma-rays, in the fireball of the Big Bang itself; but it has been redshifted, as space has stretched, all the way through the visible spectrum and out beyond the infrared, into the microwave radio bands. This weak hiss of radio noise fills the entire Universe, and is detected by radio telescopes on the ground and by instruments on board satellites such as COBE. It is this hiss that confirms that there really was a Big Bang, and tells us what the Universe was like a mere 300,000 years after the moment of its birth (the precise time corresponding to the release of this radiation marked the end of the fireball phase of the Big Bang, the almost smooth initial expansion that is known from basic physics to have lasted some 300,000 years). But before I bring that story up to date, there is another twist in the tale of the discovery that the Universe did indeed have a definite beginning. The discovery that the Universe is not eternal and unchanging, but was born and will die, did indeed come as a bombshell to astronomers in the 1920s. It might not have been quite such a shock, however, if they had been aware of, and had taken seriously, some philosophical speculations about the nature of the Universe, speculations which pre-dated the discovery of Hubble's law by fully 350 years, Thomas Digges in the 1570s. The puzzle that worried Digges, and several philosophers over the next three centuries, was why the sky is dark at night.

You might think that the answer is simple – that the Sun, our main source of light, is below the horizon at night and therefore the sky is dark. But that would be true only if there were no other sources of light in the sky. The puzzle of the dark night sky has to do with the nature of infinity – infinite space, *and* infinite time. And it was Digges who brought the notion of infinite space into cosmology.

Before Digges, the standard explanation of the starry appearance of the night sky dated back to Ptolemy, who said that the stars were little lights, attached to the inner surface of a crystal sphere, all at the same distance from the Earth. Digges asked the obvious question of what, in that case, lay outside the crystal sphere, and answered it by suggesting that the stars were not, in fact, all at the same distance from us, and were not attached to anything, but were scattered through the blackness of infinite space. This gets rid of the puzzle of the edge of the Universe – in infinite space, there is no edge. But, as Digges realized, if infinite space is full of stars, then everywhere you look you should see a star. Every single point in the sky should be shining with starlight. He got round this problem by saying that the more distant stars are simply too faint to be seen, and that the stars we do see are simply the nearer ones.

But this argument doesn't hold water. Johannes Kepler realized, in the early 1600s, that if the Universe really is infinite and really is full of stars, then the sky really should be a blaze of light. Even though distant stars are faint, in an infinite Universe filled with stars there must be an infinite number of very faint stars, and their total contribution of light would make the sky infinitely bright, no matter how faint each individual star. Kepler thought that this proved there must be an 'edge' to the Universe, a boundary where stars ceased to exist, with only blackness beyond.

It was only after Galileo's observations and Kepler's calculations established, in the seventeenth century, that the Earth is a planet orbiting the Sun that the notion of the stars as other suns became finally accepted. And it was only in the nineteenth century that the first accurate measurements of the distances to other stars were made. This led to a growing realization of just how bright the stars must really be, and encouraged further philosophical musings about the darkness of the night sky. In the eighteenth century, both Edmond Halley (who was the first person to discover that some stars move across the sky, and therefore cannot be fixed to a crystal sphere) and the Swiss astronomer Philippe Lois de Chéseaux puzzled over the problem, and in the nineteenth century it was taken up by the German Heinrich Olbers. None of them solved the riddle, which is now commonly known as Olbers' Paradox. But in the twentieth century, following Hubble's discovery of the expansion of the Universe, the answer at last became clear. The reason why the sky is dark at night is not that there is an edge to the Universe in space, but that there is an edge in *time*.

Remember that light travels at a finite speed, and spreads out from each bright star across the Universe. Light from a star five light years away takes five years to reach us; light from a galaxy 50,000,000 light years away takes 50,000,000 years to reach us. We can recast Olbers' Paradox in modern language by asking how it is possible for the sky to be dark at night if the Universe is full of galaxies so that, everywhere we look in the sky, every line of sight should end in the glow of a galaxy. Even if the Universe is 'closed' and finite in size, as I have described, if it has been around forever – for infinite time – in the same form as we see it today, then the light from all the stars and galaxies should have filled space and made the night sky blaze. The resolution of the puzzle is that in the 15 billion years since the Universe was born, there simply has not been time for all the stars in all the galaxies to shine enough to fill the Universe with light. The 'wall of darkness' that Kepler considered lies out there, in a sense, at a distance of 15 billion light years. We shall never see any light from farther away than that, because such light would have had to be emitted more than 15 billion years ago, and there were no stars to radiate light then, because there was no Universe then.

The person who first came close to the truth was, curiously enough, the novelist and poet Edgar Allan Poe, who pointed out in 1848 that by looking out into space, and therefore back in time, we may see the darkness that existed before the stars were born. But even this prescient interpretation of Olbers' Paradox was still one step away from the truth. When we look through the gaps between the bright stars and galaxies at the dark night sky itself, we are looking back to the edge of time, the moment of creation at which not just the stars but the entire *Universe* was born. And what we 'see' in those gaps is the faint hiss of radio noise that is the leftover radiation from the Big Bang itself, the oldest radiation in the Universe.

Now, at last, we can see why that radiation is so important, and we can put the COBE discoveries in their proper perspective. But as we move into the high-tech world of radio telescopes and artificial satellites, never forget that the only equipment you need to prove that the Universe was born at a definite moment in time, and is not eternal and unchanging, is your own pair of eyes looking up at the blackness of the night sky between the stars.

CHAPTER TWO

COBE in Context

The radiation that fills space between the stars and galaxies is left over from a time when everything in the Universe was squeezed into a hot fireball – the Big Bang. Winding the equations of cosmology derived from Einstein's general theory back towards the point of infinite density – the singularity – we see that there must have been a time, not just when galaxies were piled up on top of each other, but when individual stars were merged together in one huge fireball. At that time the entire Universe was as hot as the inside of a star like our Sun, and it was filled with hot radiation. All of space was filled with this hot radiation, because, remember, all of space was contained in the fireball. The radiation from the Big Bang has not expanded outwards from an initial explosion to spread through space: it has always filled what space there was available to fill, and as the space itself has expanded so has the radiation expanded with it. Expanding radiation means stretching it, which means redshifting it. What used to be radiation of very short wavelength gets stretched to longer and longer wavelengths as the Universe expands.

The most important difference between short-wavelength radiation and long-wavelength radiation is that long-wavelength radiation is cooler and has less energy. Short-wavelength X-rays and gamma-rays are much more energetic (and potentially damaging) than longer-wavelength visible light, which in turn is much more energetic than the still longer-wavelength radio waves. The radiation that started out as hot as a star in the Big Bang both stretched and cooled as the Universe expanded. Today, it has a wavelength in the microwave band, similar to the kind of radio waves used in radar systems or in a microwave oven. And it has a temperature of just under *minus* 270 degrees Celsius.

The lowest temperature it is possible to have, the absolute zero of temperature at which all thermal movement of atoms and molecules stops, is −273°C, which is defined as the zero of the absolute, or Kelvin (K), temperature scale. The cosmic microwave background radiation which fills all of space has a temperature of 2.7 K. It is radiation directly from the Big Bang itself, which we can 'see' with the aid of radio telescopes. Indeed, you can listen to the sound of the Big Bang in the privacy of your own home. Because the background radiation is literally everywhere in the Universe, some of it gets picked up, along with other forms of interference, by ordinary TV antennas. If you tune your TV set to a frequency between those of the stations broadcasting the programmes you will see the screen covered with flickering white dots, and the speaker will hiss with the noise sometimes referred to as 'static'. About 1 per cent of the incoming 'signal' causing that hiss is, in fact, cosmic microwave background radiation, direct from the Big Bang to your own front room.

This radiation had actually been discovered in the 1960s. But it was on 24 April 1992 that it became headline news, when newspapers around the world used banner headlines on their front pages to report the discovery, announced at a meeting of the American Physical Society in Washington, DC the day before, that the background radiation is not perfectly uniform. There are tiny ripples in the background radiation, corresponding to equally tiny differences in the temperature of the microwave energy reaching us from different parts of the sky.

The discovery was made by scientists based in California using instruments on board an unmanned Earth-orbiting NASA satellite called COBE, from 'Cosmic Background Explorer'. The headlines talked of the discovery of the 'Holy Grail' of cosmology, the answer to 'the riddle of the Universe', and reported 'Cosmic origin traced' while asking 'Has man mastered the universe?'. Ecstatic scientists were quoted as saying 'what we have found is evidence for the birth of the Universe' (George Smoot, Berkeley);

'it's one of the major discoveries of the century. In fact, it's one of the major discoveries of science' (Joel Primack, Santa Cruz); 'unbelievably important' (Michael Turner, Chicago); 'the missing link of cosmology' (Joe Silk, Berkeley); and, in the most over-the-top comment of all, Stephen Hawking (Cambridge) rashly claimed that the COBE results were 'the discovery of the century, if not of all time'.

Now that, surely, is pitching things a little *too* high. Even within the context of the description of the Universe that I have given in this book so far – without getting into any debate about whether some entirely different kind of discovery, like penicillin, might more justly be ranked number one in the scientific charts – the COBE results cannot be regarded as the discovery of the century. After all, what COBE found was a variation, from place to place in the sky, of the temperature of radiation that had already been discovered almost thirty years before. The discovery of the background radiation itself certainly ranks higher than the discovery of variations within it, although in the mid-sixties that discovery did not rate quite the public and media attention that the COBE discoveries received in 1992. And *the* discovery of the century, in cosmology at least, was without doubt the dramatic discovery made by Hubble, and confirmed by Einstein's equations, that the Universe is not eternal, static and unchanging, but is expanding and therefore must have been born out of a superdense state billions of years ago. But then, 'Third most important cosmological discovery of the century' does not make quite such a tasty newspaper headline as 'Most dramatic scientific discovery of all time'!

So was the story all hype, stirred up by a NASA team eager for public attention and seeking bigger and better research funding for future projects? Far from it, although the element of hype was certainly there. What COBE found really was a 'missing link' in the Big Bang theory of creation, a piece of the jigsaw puzzle that *had* to be there if the theory was correct, but which had proved extremely difficult to locate. If COBE had not found those ripples in the cosmic background radiation, the entire

Big Bang theory would have been in deep trouble, with doubt cast upon much of what astronomers had learned about the Universe since the time of Hubble, and no satisfactory alternative theory around to take its place. What COBE found is indeed a key ingredient in the Big Bang story, and one which opens the way to new cosmological ideas. As George Smoot, head of the Berkeley team that made the discovery, pointed out at that meeting in Washington, 'this *begins* [my emphasis] the golden age of cosmology. It is going to change our view of the Universe and our place within it.'

The reason why the discovery of these cosmic ripples represents the end of the beginning of the story of cosmology, rather than the beginning of the end of the story, is that now that the standard model of creation, the simple Big Bang theory itself, is confirmed as a reliable guide to how the Universe has evolved over the past 15 billion years, it becomes reasonable to ask what went before – 'Where did the Big Bang come from?' – and what goes after – 'How will the Universe die?' The answers to these questions, tentative though they must be as yet, call for a dramatic change in the way that science should view the cosmos, treating it not so much as a machine blindly following unchanging and unchangeable mechanical laws of behaviour, but as an evolving entity, adapting itself to a changing environment and competing with other, similar entities for the right to exist. But before I get too deeply into that story, it is important to set those truly remarkable and exciting COBE results in their proper context, as 'only' the third most important cosmological discovery of the twentieth century.

The first most important cosmological discovery of the century, the discovery that the Universe is expanding and did have a definite origin some 15 billion years ago, was the theme of Chapter One. But even in the 1930s, when the idea of the 'cosmic egg' or 'primordial atom' began to be discussed, nobody fully appreciated that the heat and energy of what we now call the Big Bang should have left a detectable imprint on the Universe today. Astonishingly, though, although the theorists did not realize what the measurements meant in terms of the origin and evolution of the Universe (and would not do so for another quarter of a century), observational astronomers, using the conventional techniques of spectroscopy, had already taken the temperature of the background radiation, less than ten years after Hubble's discovery that the Universe is expanding.

Spectroscopy is the key tool of astronomy, because the unique set of 'fingerprint' lines in the spectrum that corresponds to each element and each type of molecule enables astronomers to find out what the stars and the clouds of dust and gas between the stars are made of. There are stellar spectra to analyse, because stars shine brightly and produce their own spectra. But even clouds of cold material in space reveal their presence and composition through their spectra, because of the way they block out some of the light from the stars beyond them. The light is absorbed by the material those clouds are made of (their atoms and molecules) not just any old how, but at precisely the appropriate wavelengths of the spectral line patterns associated with the atoms and molecules of the clouds. Hot stuff (like a star) that radiates energy produces bright lines in a spectrum; cold stuff (like an interstellar cloud) that is absorbing light from a star beyond it produces dark lines in a spectrum.

The strength of those lines, and more subtle details of the pattern, enable astronomers to work out not just what the clouds are made of, but what their temperature is. In the 1930s astronomers identified, among other things, spectral lines corresponding to a substance known as cyanogen (a compound made from carbon and nitrogen, CN) in some interstellar clouds. As a matter of routine, they analysed the spectra to find out the temperature of those clouds, and came up with a value of roughly 2.3 K. By 1940, the figure had been published in the specialist scientific journals; by 1950, it was enshrined in the pages of standard astronomy text books. But nobody at the time, or at any time before 1964, realized the deep significance of the fact that cold clouds between the stars do not have a temperature of absolute zero, but are more than two degrees warmer than absolute zero. Today, it is 'obvious' that the clouds are kept at that temperature by the energy of the cosmic microwave background. They are being cooked in a very cool microwave oven, the Universe. But in 1940 nobody had suggested that there might be such a thing as a cosmic microwave background.

The suggestion had, though, been well and truly made before 1950, and the leading member of the team that made the suggestion came very close to seeing the significance of the interstellar cyanogen temperature a few years later. His name was George Gamow. Born in the Ukraine in 1904, Gamow fled the Soviet Union in 1933, and moved to the USA. He started out as a physicist, but became fascinated by the implications of Hubble's discovery that the Universe is expanding, and tried to use his training in physics to explain how the material the stars are made of could have been produced in the fireball in which the Universe was born. Gamow's basic idea – one thoroughly borne out by more recent calculations – was that the primordial stuff the fireball contained, when the Universe was only a few seconds old, was a hot, dense mixture of protons and electrons, interacting violently with one another and with the radiation of the fireball.

Under everyday conditions on Earth today, an electron and a proton together make up an atom of the simplest and lightest element, hydrogen. In a hydrogen atom the relatively massive proton is the nucleus and is orbited by the relatively light electron. Each proton has a mass almost two thousand times as great as an electron, and each proton has one unit of positive electric charge, while each electron has one unit of negative charge. In an atom of hydrogen, which is electrically neutral overall, the electron can be thought of as orbiting the proton, and as being held in place by the attraction between the opposite electric charges. But under the conditions of heat and pressure of the cosmic fireball in which the Universe was born, no stable atoms could exist. They would have been smashed apart in collisions, leaving electrons and protons to wander freely, in a state known as a plasma. This is exactly what happens inside the

Sun and stars today. The electrically charged particles in the plasma fireball would have interacted constantly with the electromagnetic radiation of the fireball, keeping the particles and the radiation all at the same temperature. But this kind of close linkage between particles and radiation is only possible if the particles are independent and electrically charged. When the temperature of the fireball fell below about 6000 K (roughly the temperature at the surface of the Sun today), electrons and protons could cling to one another electrically to make neutral atoms, and would no longer be able to interact freely with the radiation. At that moment, the matter and the radiation would 'decouple' and go their separate ways – the matter to form stars, planets and, eventually, us, and the radiation to fade away into a cooling background of microwave energy.

Spectroscopy reveals that the oldest stars in the Milky Way, which might reasonably be expected to contain material in more or less the form in which it emerged from the fireball of creation, are made up of about 75 per cent hydrogen and 25 per cent helium, the next simplest element after hydrogen. A helium atom has a nucleus made up of two protons and two neutrons, which are particles very similar to protons except that they have no electric charge. This nucleus, with an overall electric charge of +2, is orbited by two electrons, with a combined electric charge of −2, so that again it has no overall electric charge and does not interact freely with electromagnetic radiation. More complicated, heavier elements, such as the carbon in your body and the oxygen that you breathe, contain even more protons and neutrons in their nuclei, but always with the appropriate number of electrons orbiting outside the nuclei to keep the atoms electrically neutral. But those spectroscopic studies show that the stuff we are made of – indeed, the stuff the Earth itself is made of, and every planet that could possibly exist in the entire Universe put together – represents less than 1 per cent of the material that is found in the Universe in the form of bright stars. Stars are chiefly composed of hydrogen and helium, even today. Heavy elements are rare, and we are made of heavy elements.

Gamow and his colleague Ralph Alpher worked out how 25 per cent of the atomic stuff in the fireball of creation could have been turned into helium, before the expanding Universe got too cold for the necessary reactions to take place. If you think once again of the analogy between the expanding Universe and a loaf of raisin bread, you can imagine that by studying exactly what the finished loaf of bread is made of we could work out what ingredients had gone into the dough in the first place, and the temperature it had been baked at. This is what Gamow's team did for the Universe itself, working out the temperature of the fireball of creation and the way in which the elements in the stars had been 'baked'.

Essentially, the process consists of sticking protons and neutrons together to make nuclei of helium. In the Universe today, a neutron left on its own for a few minutes will spit out an electron, in the process known as beta decay, and become a proton. But neutrons themselves were made, under the conditions existing in the fireball, when protons and electrons were combined with one another physically, uniting to make a single particle and cancelling out their electric charges in a process known (logically enough) as inverse beta decay. Given protons and neutrons, Gamow and Alpher could explain where 99 per cent of the atomic stuff in the Universe came from, provided the temperature of the fireball itself had a certain value. If the fireball were very much hotter, or very much cooler, than the value they calculated, then the amount of helium produced would be significantly different from the 25 per cent measured in old stars. They couldn't explain where the other 1 per cent (including the stuff we are made of) came from (Fred Hoyle would do that, as we shall see, in the 1950s); but it was an impressive achievement none the less, carried out less than twenty years after Hubble's discovery that the Universe is expanding.

Gamow was an inveterate practical joker, and when the time came to publish the results of the work he and Alpher had done, his sense of humour overcame him. Deciding that the names Alpher and Gamow on the paper 'seemed unfair to the Greek alphabet', he added the name of another physicist, Hans Bethe, before submitting the paper for publication. The first Bethe knew about this was when the paper appeared in print in the *Physical Review*. To Gamow's huge delight, and entirely by coincidence, the date of that issue of the journal was 1 April 1948; and to this day the article is known to astronomers as the 'alpha, beta, gamma paper'. The joke is even better than it seems at first sight, since an alpha particle is another name for a helium nucleus, a beta-ray is another name for an electron (hence 'inverse beta decay'), and a gamma-ray is another name for a pulse of energetic radiation.

Later in 1948 Alpher and Robert Herman, another young researcher working with Gamow, extended the idea of a hot fireball of creation in which helium had been created by the fusion of hydrogen nuclei. In order to make the amount of helium left over from the fireball fit the observations of the composition of old stars, they had to set the temperature of the fireball to within a very narrow range of possibilities. It turns out that it is very simple to calculate the temperature of the radiation from the fireball at any time after the fireball era, as the Universe has expanded and cooled. In effect, you take the temperature of the Universe when it was only one second old (about 10,000,000,000 K, according to the calculations of helium synthesis) and divide it by the square root of the age of the Universe in seconds. Alpher and Herman calculated that the temperature of the leftover radiation from the cosmic fireball would today be about 5 K. The same kind of calculation tells us that the era of decoupling, when matter and radiation first went their separate ways and the temperature of the entire Universe had fallen to a mere 6000 K (as hot as the surface of the Sun) would have occurred just 300,000 years after the birth of the Universe. Because the cosmic background radiation last interacted with matter at that time, Gamow's team knew that if this radiation with a temperature of around 5 K could ever be detected and analysed, it would reveal a picture of what the Universe was like at that time. But Alpher, Herman and Gamow did not think that there was any way to measure the existence of such low-temperature radiation filling all of space, and none of them made the connection with the already known temperature of cyanogen in interstellar clouds.

Tantalizingly, at the same time that Alpher and Gamow were making their calculations, another young physicist was actually making observations of background radiation from the sky. Robert Dicke, who was then working at Princeton, had been working on radar during the Second World War, and used his expertise to develop an instrument, called a Dicke radiometer, which could measure microwave radiation from space. He pointed one of these instruments at the sky to see if there was any trace of microwave radiation coming from distant galaxies (never dreaming for one moment that such radiation might come from empty space itself), and found a weak hiss of radio noise corresponding to radiation with a temperature somewhere below 20 K, which was as much as the instruments of the day could reveal. The discovery was actually published in the *Physical Review* in 1946, but nobody made the connection with the work by Gamow's group. Gamow knew that the Universe ought to be full of cold microwave radiation, but did not realize that the technology existed to measure the radiation; Dicke had the technology to measure the radiation, but did not realize that the radiation was important enough to make a big effort to take its exact temperature.

It was almost as if the cosmic microwave background radiation was trying to be discovered, but human astronomers were proving too stupid to see it. It tried again, and failed, to be discovered in 1956.

By then, Fred Hoyle was firmly established as a leading proponent of the Steady State theory, mentioned in the Prologue, and had stuck the label 'Big Bang' on to the rival theory of cosmology, intending it as a term of derision. George Gamow was an equally ardent proponent of the Big Bang theory, and he and other supporters of the notion had eagerly taken up Hoyle's term of derision and claimed it as their own. Gamow and Hoyle remained on good terms personally, in spite of the professional rivalry in the 1950s between the two cosmological camps. In an article published in *New Scientist* in 1981, Hoyle recounted how the two friends just missed 'discovering' the cosmic background radiation while cruising around La Jolla, in California, in a brand new Cadillac convertible in the summer of 1956. At that time, remember, Gamow thought that the temperature of the background radiation ought to be at least 5 K. Hoyle thought that there should be no background radiation at all, since he did not believe there had been a Big Bang fireball. As Hoyle tells it:

There were times when George and I would go off for a discussion by ourselves. I recall George driving me around in the white Cadillac, explaining his conviction that the Universe must have a microwave background, and I recall my telling George that it was impossible for the Universe to have a microwave background with a temperature as high as he was claiming, because observations of the CH and CN radicals by Andrew McKellar had set an upper limit of 3 K for any such background. Whether it was the too-great comfort of the Cadillac, or because George wanted a temperature higher than 3 K whereas I wanted a temperature of zero K, we missed the chance.

By the early 1960s, however, the microwave background radiation was crying out to be discovered. Theorists in both England and the Soviet Union had started to draw attention to the possibility of measuring such radiation, and at Princeton a team headed by Robert Dicke (who had forgotten all about his measurements of radiation from the sky in 1946!) was busily building a new version of the Dicke radiometer to make the necessary observations. The final twist in the tale was, though, still to come. The second most important cosmological discovery of the century, the discovery of the background radiation itself, was actually made not by the astronomers who were aware of the prediction that such radiation might exist, and were actively seeking to find it, but by two young researchers, working at the Bell Laboratories in New Jersey, who had not the faintest idea what it was they had discovered until Dicke told them.

The Bell Telephone Company (as it then was) had a long tradition of sponsoring pure scientific research, as well as research directed more obviously at ways to improve communications around the world. As part of the development of a global communications network, Bell had built a kind of radio receiver known as a horn antenna, on Crawford Hill near Holmdel in New Jersey. This antenna had been used in some of the first experiments in bouncing radio signals around the world with the aid of artificial satellites. Today, such a trick is routine; radio, TV and telephone communications are regularly squirted up to satellites high above the Earth, which pick up the signal, amplify it and re-broadcast it back down to Earth. The signals are so powerful that many ordinary households have an antenna capable of picking up the satellite TV signals.

But in the early 1960s, things were different. The satellites used in the early experiments were simply large balloons, inflated automatically in Earth orbit and covered with a film of metallic material to reflect radio waves. They were called Echo satellites, because all they did was reflect back to Earth a radio signal beamed up to them from a ground station. This meant that you could, for example, send a signal up to the balloon from California, and have it bounced back down to a receiver in New Jersey. Because the Echo satellites did not actively pick up the radio signals, amplify them and re-broadcast them, however, the signal that returned to Earth after being bounced off the metallized balloon was very weak. So Bell needed a large and efficient antenna and a sensitive amplifier system in order to pick up the signals that came back down to Earth. The Crawford Hill antenna was purpose-built for the job, and was designed to be particularly sensitive to microwave radiation in the waveband used in the Echo experiments. It looked rather like a huge version of an old-fashioned ear trumpet, with a rectangular opening, six metres (twenty feet) across at one end to catch the microwaves, and a tapering, hollow arm on the side to guide them into the amplifier.

When the series of experiments with the Echo satellites was

over, Bell made the horn antenna available for radio astronomy. The principle of a radio telescope is exactly the same as that of a communications antenna and receiver – it catches a radio signal and amplifies it. (Astronomers use the term 'signal', by the way, to refer to any radio noise from space; there is no suggestion that the use of the word implies that there is any intelligence at work behind the signal, any more than there is intelligence at work behind the signal of light that we receive from the Sun.) In radio astronomy, though, the signals being investigated come from the depths of space, not from satellites in Earth orbit.

Penzias and Wilson were interested in using the horn antenna on Crawford Hill to measure the radio noise coming from the Milky Way Galaxy itself, in directions well away from the heart of the Milky Way. They expected to find radio waves coming not from any individual active stars, but from the spread-out contribution of individual hydrogen atoms in space, interacting with the magnetic field of the Galaxy. By analysing such signals it would be possible to map out the distribution of hydrogen gas between the stars (this has, in fact, now been done). But Penzias and Wilson ran into a snag. In the spring of 1964, when they had completed converting the horn antenna into a radio telescope and started making their observations, they found a much stronger signal than they had expected – radio noise with a wavelength of just over 7 cm coming evenly from all directions on the sky.

As the weeks and months wore on, the Bell team found no change in the signal with the time of day or with the seasons. Since the Earth both rotates on its axis and moves around the Sun, so that a ground-based antenna has a constantly shifting orientation, this showed that the signal must be genuinely uniform. It was so strong, though, that if it was indeed coming from the interstellar gas in our Galaxy then the Galaxy itself would shine like a radio beacon in the Universe. There is no reason to think that our Milky Way Galaxy is special, and if our Galaxy could shine so brightly at radio frequencies, other galaxies should do the same. Yet those other galaxies – the Andromeda Galaxy, for example – show no sign of such strong radio emission.

Although weak by the standards of any conventional TV signal, the signal was, in fact, so much stronger than anything they had expected to find that Penzias and Wilson thought at first that there must be something wrong with their equipment. But every test showed that the system was working perfectly. They even worried that a liberal coating of pigeon droppings that had built up inside the horn antenna might be affecting its electrical properties, and the whole horn was painstakingly dismantled, cleaned and re-assembled, but there was no change in the signal. Penzias and Wilson really were detecting microwave signals from space, with a temperature of just under 3 K. They were completely baffled.

In December 1964 Penzias mentioned the puzzling signal in a conversation with Bernard Burke, of the Massachusetts Institute of Technology (MIT). The following month, Burke called Penzias to let him know that yet another astronomer, Ken Turner of the Carnegie Institution in Washington, DC, had told Burke about a talk given by a Princeton researcher, Jim Peebles, in which he had predicted that the Universe should be filled with electromagnetic radiation (radio waves) with a temperature somewhere below 10 K. Peebles was a member of Dicke's team at Princeton, just 50 km (30 miles) away from the Bell Laboratories. A phone call from Penzias to Dicke soon brought all four members of the Princeton team over to Holmdel to compare notes with the Bell team. At last, twenty years later than might have been, the theory of the Big Bang and the observations of a cosmic microwave background radiation were put together.

The news spread rapidly among the astronomical community, and the two teams from New Jersey published papers alongside each other in the *Astrophysical Journal*. Outside the astronomical community, the discovery of the background radiation received only modest attention. But to astronomers in general, and to cosmologists in particular, the world – the Universe – would never be the same again. As more and more observations were carried out at different radio observatories, confirming that the radiation discovered by Penzias and Wilson really was the background radiation required by the Big Bang theory, and refining its temperature to 2.7 K, cosmologists had to confront the incredible truth that the equations that they scribbled on their blackboards and played with in their computers really were describing the birth of the Universe itself – that there really had been a birth of the Universe, and that the combination of measurements of the temperature of the background radiation and of the amount of helium in old stars really could reveal what the Universe was like during the fireball era, just minutes, seconds even, after it was born. The impact on astronomers of the discovery of the cosmic microwave background radiation was as dramatic as the impact on the world at large of the discovery by the COBE team, almost thirty years later, of the ripples in that radiation.

Today it is hard to appreciate what a shattering discovery the identification of the background radiation was. The fact is, that from the time of Hubble right up to the discovery of the background radiation (made in 1964 but announced in 1965), cosmology had been no more than a mathematical game. The cosmologists could develop equations to describe various forms of Big Bang 'model' mathematically, or to describe rival ideas such as the Steady State hypothesis, but (with the possible exception of Gamow's team) nobody really believed, at gut level, that the models held any ultimate truth. It was like a game of cosmic chess, in which researchers tried to find ways to explain particular observations (such as the fact that the Universe expands) more out of interest in the mathematics of the equations than in the hope of finding out how the Universe *really* began. Physicist Steven Weinberg expressed it succinctly in his book *The First Three Minutes*, published in 1976. 'It was', he said, 'extraordinarily difficult for physicists to take seriously *any* theory of the early universe' in those days. 'Our mistake is not that we take our theories too seriously, but that we do not take them seriously enough. It is always hard to realize that these numbers and equations that we play with at our desks have something to do with the real world.' Elaborating on this theme, he paid tribute to the pioneering work of Gamow and his colleagues (who were bitterly hurt when the 'discovery' papers in the *Astrophysical Journal* made no reference to their pioneering efforts, but were later given full acknowledgement):

> Gamow, Alpher and Herman deserve tremendous credit above all for being willing to take the early universe seriously, for working out what known physical laws have to say about the first three minutes. Yet even they did not take the final step, to convince radio astronomers that they ought to look for a microwave radiation background. The most important thing accomplished by the ultimate discovery of the 3 K radiation background in 1965 was to force us all to take seriously the idea that there *was* an early universe.

The power of this new insight became apparent within a couple of years of the announcement of the discovery of the background radiation, when researchers were able to use the precise information about the Big Bang fireball provided by the measurements of the precise temperature of the radiation (with the aid of improved information about the kind of nuclear reactions that convert hydrogen into helium and heavier elements) to produce the best description yet of the way primordial ingredients had been 'cooked' in the Big Bang itself.

It is ironic that one of the leading lights in this work in the mid-1960s was Fred Hoyle, who ten years previously had been cruising around La Jolla in a white Cadillac convertible trying to persuade George Gamow that there had been no Big Bang. But Hoyle had long had a special interest in the origin of the elements – indeed, that was why he was in California in 1956 – and the study of Big Bang 'nucleosynthesis', as it is called, followed on naturally from this.

The story actually goes back to Gamow's success in explaining where 99 per cent of the stuff of the visible Universe (the hydrogen and helium) came from, and his failure to account for the origin of the other 1 per cent (which includes us). Starting out in 1946, Hoyle had been following up a different line of attack, considering how heavier and more complex elements might have been built up from hydrogen by nucleosynthesis inside stars. Nucleosynthesis stops in the Big Bang once the Universe has expanded and cooled to the point where light nuclei can no longer smash together hard enough to make them 'stick' to one another and form more complex nuclei. So the era of Big Bang nucleosynthesis ends within a few minutes of the birth of the Universe. The temperature and pressure inside a star are never as extreme as the temperatures and pressures during the first moments of the birth of the Universe, but a star has the great advantage that it sits around in space for a very long time – in some cases, billions of years – and during all that time a steady trickle of nucleosynthesis may be going on in its interior.

Hoyle worked out how the more complex and heavier elements such as carbon and oxygen might be put together from lighter elements such as helium inside stars. His theory could be tested only by measurements of the way light nuclei interact when they are smashed together in particle accelerators here on Earth. Those measurements were carried out by researchers at the California Institute of Technology (Caltech), where Hoyle's close friend and colleague Willy Fowler worked (which is why Hoyle was visiting California in the summer of 1956). The experiments showed that Hoyle was right: that a star which started out made up of about three-quarters hydrogen and a quarter helium, with just a tiny trace of a few other elements such as lithium, would steadily manufacture heavier elements in its interior in just the right way to account for our own existence. Those heavier elements would be scattered through space by stars blowing up at the end of their lives, puffing material out into space where it could eventually form new stars and planets. This was an enormously important discovery, but it left one question unresolved. Where had the original hydrogen, helium and traces of things like lithium come from to make the first generation of stars? It looked as if Gamow's ideas about Big Bang nucleosynthesis might be needed after all.

In the 1960s, the Caltech team turned their attention to the kind of nucleosynthesis reactions that might have taken place in the Big Bang, updating Gamow's work and improving it with the aid of measurements of more than a hundred nuclear reaction processes in their laboratory, and the new-found knowledge of the exact temperature of the cosmic fireball, revealed by the measurements of the background radiation. Just as you can work out the temperature of the radiation today if you can estimate what the temperature of the fireball ought to have been, so can you work out precisely what the temperature of the fireball was if you know the precise temperature of the radiation today. This in turn makes it possible to calculate exactly how the stuff of the Universe should have been cooked in the Big Bang fireball. In a *tour de force* paper published in 1967, Fowler, Hoyle and a young researcher named Robert Wagoner established that the proportions of hydrogen, helium and lithium coming out of the Big Bang did indeed precisely match the proportions found in old stars. Combined with the earlier work on stellar nucleosynthesis, this showed how everything in the visible Universe was created out of primordial protons and electrons in a two-stage process: first, light elements were cooked in the Big Bang, then heavier elements were cooked more slowly inside stars.

The Wagoner, Fowler and Hoyle paper depended crucially on the evidence for the hot Big Bang from the studies of the cosmic microwave background. Astronomers who had been struggling to come to terms with the discovery saw any doubts about the interpretation of the microwave background swept away. One of the pioneers of cosmology, Sir William McCrea (who was born in the same year as Gamow), says that 'it was this paper that caused many physicists to accept hot Big Bang cosmology as serious quantitative science'.

In the 1970s and 1980s, this serious quantitative science became the most important game in town for physicists. The background radiation itself was measured in ever more precise detail using instruments on the ground and on balloons, high-flying aircraft, rockets and satellites, pinning it down ever more confidently as the 'echo of creation'. The calculations of nucleosynthesis, both in the Big Bang and in the stars, were improved and refined, matched to the background radiation, and used to provide an even more precise and detailed description of the early stages of the Big Bang. And interest in cosmology spread outside the ranks of cosmologists, and even outside the ranks of astronomers, as mainstream physicists such as Steven Weinberg saw that one of the best ways to test their theories about how particles interact under extreme conditions was to see how well those theories matched up to the Big Bang description of the birth of the Universe.

But while all this work was going on, there were two clouds on the horizon. Initially they were quite small, but they grew to assume more importance. The first problem was that, as time went by and more observations were made, it seemed that the background radiation was simply *too* smooth. Remember that this radiation provides us with a kind of radio picture of what the Universe was like at the time when matter and radiation decoupled, some 300,000 years after the birth of the Universe. Up to then, matter and radiation had been tightly linked with each other. So the smoothness of the background radiation tells us that all the matter in the Universe (the hydrogen and helium) was distributed very, very smoothly at that time. And yet, today the hydrogen and helium is clumped together in stars, which are gathered together in galaxies, and even galaxies tend to group together in clusters separated by vast tracts of seemingly empty space. How could a very, very smooth gruel of hydrogen and helium, cooling in an expanding Universe, have formed clumps as big as galaxies and clusters of galaxies in the time available? Computer simulations of the way galaxies would have grouped together under the influence of their own gravity said that it simply could not – that the Universe is far too young (given its initial smoothness) for galaxies and stars (and, by implication, ourselves) to exist.

The other puzzle concerned the way galaxies move through space. Although galaxies are far too far away from us for their movement across the sky to be perceptible, the speed at which they are moving can be inferred from spectroscopy, using the

Doppler effect. In a cluster of galaxies, all about the same distance from us, each galaxy will have about the same cosmological redshift, caused by the expansion of space. So any differences in the redshifts of the galaxies in a cluster must be due to the Doppler effect of their actual motion through space, as they orbit around one another. The puzzle is that these speeds seemed to be too great for the galaxies to be held in orbit around one another by gravity. Clusters should not exist, even if individual galaxies had had time to form – the galaxies ought to be flying apart from one another, unless there is some extra, unseen dark matter present in the cluster, holding the visible, bright galaxies in its gravitational grip. But a lot of this dark stuff would be needed, and the ever-improving Big Bang calculations said that just about all of the hydrogen and helium that could have been produced in the Big Bang was accounted for by the visible stars and galaxies. If dark stuff exists, it would have to be some different form of matter entirely, never found on Earth or in the bright stars.

The two puzzles were resolved in the 1980s by a beautiful theory known as inflation, combined with a suggestion that the space between the galaxies contains huge quantities of stuff known as dark matter. Inflation says that during the first split-second of the birth of the Universe, a tiny seed containing all the mass and energy in the observable Universe was blown up from a size smaller than that of an individual proton to about that of a basketball. This very rapid inflation would have smoothed out any irregularities in the distribution of matter and energy (and even any irregularities in the structure of space itself), in the same way that the wrinkles on the skin of a prune are smoothed out when it is soaked in water and expands. Then, at the end of the inflationary era the energy of the rapid expansion would have been turned into mass-energy, as the expansion slowed down to the more sedate pace that we see today. All of the now-standard activity of nucleosynthesis and interactions between radiation and matter in the cosmic fireball would have taken place in the very smooth Universe as it expanded from the size of a basketball over the next 300,000 years or so, leaving a very smooth background radiation to this day. But inflation also requires that a lot of the energy that powered the initial expansion be turned into matter (in line with $E = mc^2$) and mixed with the protons and electrons of the standard Big Bang model. That extra matter would *not* be in the form of protons and electrons, but something else entirely. Unlike the protons and electrons, it would have no electric charge, and so it would not be kept smoothed out by the influence of the radiation, and it could begin to clump together to make irregularities in the expanding Universe even before everyday matter and radiation decoupled. At the time of decoupling, according to this new idea, there ought to have been ten or a hundred times as much matter in the form of this dark stuff as there was in the form of hydrogen and helium, already formed into irregularities that could attract the hydrogen and helium by gravity. It pulled clouds of gas together to make stars, galaxies and clusters of galaxies, and is still holding those clusters together, in spite of the large orbital velocities of many individual galaxies in clusters, to this day.

The combination of inflation and dark matter explained both why the background radiation was so smooth 300,000 years after the birth of the Universe, and how there had been time for galaxies to form out of this smoothness over the next few billion years. It also made one crucial prediction. If the whole package of ideas – inflation, dark matter and the standard hot Big Bang theory – were correct, there ought to be irregularities in the background radiation corresponding to the disturbances produced as the hydrogen and helium gas began to be tugged in towards the clumps of dark matter just at the end of the fireball era. The irregularities would show up as tiny differences in the temperature of the background radiation, ripples amounting to deviations of only a few tenths of a millionth of a degree from the average temperature, from different parts of the sky. Because the background radiation last interacted with matter 300,000 years after the birth of the Universe, 15 billion years ago, these deviations would in effect be ripples in time, echoing down billions of years to tell us how matter was first distributed. And they were far too small to have been detected by any radio telescope used to study the background radiation before the end of the 1980s. Which is where COBE came in.

Space probes always have a very long gestation time, with years spent planning, designing and building the instruments and the satellite itself before it ever gets off the ground. COBE suffered more delays than most. The idea came from a young researcher, John Mather, in 1974, when he was a graduate student at Columbia University. The proposal, made just ten years after the discovery of the cosmic microwave background, was for a fairly modest spacecraft to survey the background radiation from above the Earth's atmosphere. Even with fairly small and simple radio telescopes on board the satellite, it would, Mather pointed out to NASA, have great advantages over ground-based instruments.

Radio astronomers have no trouble observing the background radiation from the ground, and even in taking its temperature to an accuracy of a hundredth of a degree; but they have severe difficulties when it comes to investigating the kind of tiny ripples in the radiation that would have to be there to explain the existence of galaxies and ourselves. When they make observations at relatively long wavelengths (a few centimetres or so) the weak hiss of radio noise from the background itself is mixed up with radio noise from the Milky Way (including the kind of radiation Penzias and Wilson had originally been looking for, back in 1964). The patchiness of this Milky Way signal, with some regions hotter and brighter than others, masks any small variations in the background signal from different parts of the sky. At shorter wavelengths the interference from the Milky Way is much less, but the radio astronomers then run into problems with water vapour in the Earth's atmosphere, which absorbs some of the incoming microwave background. This is why over the years different groups of radio astronomers have tried to tilt the odds in their favour by making observations from instruments flown on balloons, aircraft and rockets, and from observatories high on mountain tops (above much of the water vapour in the atmosphere) or located in the cold, dry air of Antarctica. But the best answer to all the problems was clearly to put a microwave detector (in effect, a short-wave radio set hooked up to a small horn antenna) on board a satellite. NASA backed Mather's idea, which became the Cosmic Background Explorer, and drew in many different researchers to work on the project, including David Wilkinson, who had been a member of Dicke's team that had so nearly made the discovery of the background radiation a decade before. Mather himself became Chief Scientist on the project.

In the second half of the 1970s and into the early 1980s, the COBE concept evolved into a drum-shaped spacecraft, 3 metres (10 feet) across and weighing 5 tonnes, tailored to fit into the cargo bay of a Space Shuttle. It was ready for launch in January 1986, and booked on a Shuttle flight when the shuttle *Challenger* was destroyed in an explosion shortly after takeoff.

After the disaster, with Shuttle flights suspended, many satellite projects had to find other launchers. In the case of COBE, this meant adapting the satellite to ride into space on top of a Delta rocket. Unfortunately, in order to fit inside the nose cone of the Delta launcher, the satellite could be no more than 2.4 metres (just under 8 feet) in diameter. And in order to get it into the right orbit on a Delta launcher, its weight had to be cut by half.

The satellite had to be taken apart and rebuilt to fit the launcher, with some equipment thrown out to save weight. But there was one advantage in using the Delta rocket. COBE had to go into an orbit 900 km (550 miles) up, and because the Shuttle only flies into a much lower orbit, the original design had included a rocket to carry the satellite higher after it had been released from the Shuttle's payload bay. Because the Delta launcher could take the satellite all the way to 900 km, the extra rocket motor could be left out of the redesign, saving a whole tonne. Changes in the solar cells used to power the satellite and in the shielding used to protect the instruments from cosmic rays completed the weight saving, and the redesigned and rebuilt COBE satellite was successfully placed into the required orbit on 18 November 1989.

Everything worked perfectly. Almost immediately, COBE was measuring the overall temperature of the microwave background more accurately than ever before, as 2.735 K. But everyone was on tenterhooks, waiting to know whether there was any sign of the tiny variations in temperature from place to place that would reveal the birth pangs of clusters of galaxies. They had to curb their impatience a little longer, because in order to answer that question COBE had to scan the entire sky, and at three different wavelengths. The observations needed to construct a temperature 'map' of the sky took more than a year to complete, and involved 70 million separate measurements from each of the three separate radiometers. It then took months for the team of thirty-four scientists to process the raw data and carry out the computer aided analysis that revealed, at last, the presence of the ripples. The bright spots of the background radiation on the sky are just 30 millionths of a degree hotter than average, while the cool spots are just 30 millionths of a degree cooler than average.

But there is still more to the story. The COBE data also show that fluctuations of just the same amount, 30 millionths of a degree, are seen on all scales. The smallest patch of sky analysed by the COBE detectors covers about fourteen times the diameter of the full Moon as seen from Earth, and the same variations in temperature are found on patches from this size up to one-quarter of the entire sky. Big hot spots (or cold spots) are no hotter (or colder) than average than are small hot spots (or cold spots). This 'scale invariant' pattern of fluctuations is exactly what had been predicted by the theory of inflation, which says that the structure we see in the Universe today was imprinted by quantum processes when the entire visible Universe was much smaller than an individual proton. And there is another bizarre aspect to this discovery, one that caused enormous confusion to some of the more perceptive readers of the enthusiastic newspaper stories that greeted the announcement of the COBE results.

The biggest of those ripples in the cosmic background radiation, stretching across a quarter of the sky, correspond to tenuous clouds of matter spread across distances of around 500 million light years. But remember that these are ripples in time, from an era when the Universe was only 300,000 years old. Since nothing can travel faster than light, and one light year is the distance light travels in one year, then how, many puzzled readers asked, can structures 500 million light years across exist in a Universe only 300,000 years old? The answer is that while nothing can travel *through space* faster than light, the expansion of the Universe, as I explained in Chapter One is caused not by galaxies moving through space, but by space itself expanding and carrying galaxies apart. During the first 300,000 years of the life of the Universe, the space in which the ripples now detected by COBE are embedded was expanding 'faster than light', so that a tiny fireball of mass and energy could indeed expand to more than 500 million light years across in just 300,000 years, while tiny quantum fluctuations within that fireball were stretched to become the ripples in the background radiation that we can now detect.

The ripples in time themselves were set up by slight concentrations of gas, everyday hydrogen and helium, beginning to be sucked in towards the massive concentrations of dark matter that underlay the ripples. These concentrations of dark matter are the 'seeds' from which galaxies and clusters of galaxies grew. The dark matter is sometimes referred to by cosmologists and particle physicists as 'exotic' matter, because it must be in the form of neutral particles quite unlike the protons, neutrons and electrons that together make up the atoms of every element here on Earth, and with which we are familiar. But this is something of a misnomer, since the 'exotic' matter actually makes up 99 per cent of the mass of the Universe. It is our everyday stuff that is really the exotic minority. Indeed, we are doubly exotic, since while as little as 1 per cent of the mass of the Universe may be in the form of protons, electrons and neutrons, all but 1 per cent of that everyday matter is in the form of hydrogen and helium in stars. The Earth on which we live, the air that we breathe and our own bodies are made up of a kind of matter that represents just 1 per cent of 1 per cent of the mass of the Universe.

The presence of so much dark matter in the Universe, whatever its exact form, also has another important implication. Although the Universe is still expanding today, it is doing so more slowly than during the hectic early years of the Big Bang. The reason is that the gravity of all the matter in the Universe is holding back the expansion, slowing the rate at which space is stretching. A useful analogy is with the flight of a ball thrown straight up into the air. It starts off moving rapidly, slows down, comes to a halt at the top of its flight and then falls back to Earth with ever-increasing speed. If you could throw a ball hard enough, though, it would escape entirely from the Earth's gravitational pull, and fly off freely into space. Astronomers used to think that the expansion of the Universe was like that, and would continue forever, because they knew there was not enough matter in all the bright stars and galaxies to provide a strong enough gravitational grip to halt the expansion. When you add in the influence of the dark stuff, however, it now seems that there *is* enough matter for its gravitational influence eventually to halt the expansion.

One day, tens of billions of years from now, the galaxies will stop moving apart. Then, slowly at first but with increasing speed, space will contract, pulling everything back together, after further tens of billions of years, into a hot fireball – a mirror-image of the Big Bang, known as the Big Crunch. In terms of curved space and Einstein's general theory of relativity, this means that space is indeed gently curved completely around itself, so that the entire Universe is 'closed' and self-contained, the equivalent in three dimensions of the curved, closed two-dimensional surface of the Earth itself, and also, as I shall discuss in Part Four of this book, exactly equivalent to a black hole, with us on the inside.

The Universe, COBE tells us, was indeed born in a Big Bang, is developing and ageing as time passes, and will one day die. The temptation to draw an analogy with a living system is well nigh irresistible. I hope, later in this book, to persuade you that this may in fact be more than an analogy. But before I can do this, it seems like a good idea to decide just what we mean by the term 'life' itself.

PART TWO

What is Life?

CHAPTER THREE

Life Itself

J. B. S. Haldane once remarked that even the Archbishop of Canterbury is 65 per cent water. The point he was making is that living things are composed of ordinary chemical substances – atoms and molecules – which, like water, are not in themselves necessarily living. What is it that distinguishes some arrangements of atoms and molecules from others, and makes those arrangements 'alive', while a glass of water is not?

Trying to write down a definition of life is rather like trying to write down a definition of time. Just as we all know what time is, until someone asks us to explain it, so we all know what life is, until someone asks us to explain it. The most obvious and standard definition of life is that living things have the ability to reproduce themselves, to grow, and to respond to changes in their environment. A lump of rock clearly is not alive, since it does not reproduce or grow, or have enough sense to come indoors out of the rain. But this is an unsatisfactory definition of life, because we can all think of things that are clearly not alive, but which satisfy some or all of these criteria. A crystal, perhaps of a substance as common as salt, will grow if it is suspended in a suitable solution; a flame seems both to grow and to reproduce; a soap bubble floating in the air can seem to be responding to external stimuli, retreating from a hand held out to touch it.

A better way to define life would surely be to introduce the notion of complexity – the Archbishop of Canterbury is certainly a more complex structure, chemically speaking, than a glass of water; and, for all their life-mimicking properties, a crystal of salt, a flame or a soap bubble are all simple chemical structures. But we encounter another, probably more profound blurring of the distinction between living and non-living when we try to find just where on the scale of things we can draw the line between the two. A human being is clearly a living creature with complex chemistry; a glass of water has a very simple chemical structure and is clearly not alive. But you do not need quite as much complexity as a human being in order to be absolutely certain that you have a living thing. Creatures such as ourselves are made up of many individual cells, and each microscopic cell possesses all the usual attributes of life, including the ability to reproduce and to respond to the outside world. Haldane pointed out, in 1929, that there are about as many cells in a human being as there are atoms in a cell. Individual atoms are clearly not alive, and, said Haldane, 'the line between living and dead matter is therefore somewhere between a cell and an atom'.

Somewhere in that range of sizes there is a kind of entity known as a bacteriophage (or phage, for short) which is a type of virus. Bacteria are single cells which live out their life cycles without recourse to grouping together to form a more complex organism. Phages are smaller than cells, but quite complex structures in chemical terms, that do not seem to be alive when they are isolated, but which can hijack the reproductive chemistry of a bacterium, using the cell's machinery to manufacture more copies of the phage, which can each then go on to infect other bacteria. A phage cannot reproduce itself unaided, and it survives, as Haldane expressed it, 'heating and other insults which kill the majority of organisms'. If something cannot be killed, our everyday experience would suggest that it was never alive in the first place. Yet a phage can reproduce by taking over bacteria. Is a phage living or non-living?

The most sensible answer is that phage cannot be regarded as a living substance in itself, but that the combination of phage plus bacteria very definitely is alive (even though the hijacking of the bacterium's biochemistry by the phage inevitably results in the death of the bacterium itself!). If you are looking for the level of complexity at which life 'emerges', the line should be drawn somewhere between a phage and a bacterium. But there is much deeper significance here, more than just an example which allows us to set an arbitrary dividing line between the living world and the non-living world. The whole point about the debate, the only reason that there is any doubt about the status of the phage, is that the phage cannot be treated in isolation. Although the example is extreme, it makes the point that life always involves an interaction with the surrounding environment. Living things must constantly take in raw materials from the outside world, and use them to build up the complex chemicals which are a feature of living things. Phage may take this to extremes, but you and I do it all the time – we eat, drink and breathe. And we too need as input to our living bodies some things which are produced only by other living creatures – not just the essential amino acids that are part of the food we eat (food which was itself once alive), but even the oxygen in the air that we breathe, oxygen that was put there by the action of other living creatures.

Paul Davies and I have argued, in our book *The Matter Myth*, that this complex interdependence of living things is an essential requirement of the life process, and that it is nonsense to try to think of any individual organism in isolation as 'alive'. Isolate a human being, or any other living thing, from its environment, and within a very short time that organism would be dead. The only genuinely living *system* that we know of from our own direct experience is the entire biosphere of the Earth, and it is doubtful whether any single individual organism from that biosphere would continue to 'live' if it were transported to the arid, airless surface of the Moon. You and I depend, literally for our lives, on our surroundings and our interactions with other living things, every bit as much as a phage depends on a bacterium for its life.

I shall take a closer look at this idea of the interaction of all living things on Earth to form a single 'super organism' – the concept of Gaia, associated with the work of Jim Lovelock – in Chapter Five. But the key point to take on board here is that, although we can safely say that the minimum level of complexity dividing living things from non-living things lies somewhere between an atom and a single cell, and probably between a phage and a bacterium, *there is no known upper limit to the size or complexity of living systems.* Opponents of the Gaia hypothesis sometimes argue that the entire biosphere of the planet cannot be alive in the conventional sense of the word, because it does not reproduce itself, by seeding the Universe at large with other biospheres. And it certainly takes a great leap of the imagination to envisage structures on still larger scales as living. But the Universe itself need not be circumscribed by the limited power of human imagination. There is even serious scientific reason to suspect not only that our biosphere – Gaia – is capable of reproduction, but that life on Earth may itself be the product of a 'seeding' process, so that our single biosphere is neither the beginning nor the end of the story of life within the Universe. In order to understand this, however, even though the story of this book is really one of the complexity and richness of structure associated with the evolution of living things, we have first to take a brief trip down to the simple 'nuts and bolts' level of the chemistry of the atoms and molecules of which life itself (as we know it) is composed.

Chemistry is the science of how atoms stick together to form molecules. An element – such as hydrogen or helium – is a substance made up of atoms of a single type. Compounds contain atoms from two or more different elements, held together by electric forces. Inside a star, the nuclei of different elements are built up by fusing together the basic building blocks (protons and helium nuclei) left over from the Big Bang; but under the conditions of heat and pressure that prevail inside a star these positively charged nuclei roam freely in a sea of negatively charged electrons, a state known as a plasma. Under the conditions of temperature and pressure that prevail on the Earth, each atomic nucleus has its quota of electrons strongly associated with it to form an electrically neutral atom. The way in which nuclei and electrons are associated, however, defies common sense.

Electrons have negative charge and nuclei have positive charge, so you might think that, since opposite charges attract, they ought to 'get together' completely, with the electrons falling in to the nuclei to cancel out their charge. In fact, the electrons maintain a respectable distance from the nuclei to which they belong, forming a ball of negative charge surrounding the inner kernel of the positively charged nucleus.

Even worse (as far as common sense is concerned), even the single electron of the simplest atom (hydrogen) 'surrounds' its nucleus (a single proton). The idea seems bizarre in terms of the behaviour of particles in the everyday world, not least since an electron is much smaller than a proton. It is like asking you to accept that a golf ball can 'surround' a basketball, all the while keeping far enough out from the basketball to ensure that the two objects never touch each other. The explanation is that, on the scale of electrons and protons, 'particles' do not behave in the same way as objects like golf balls do on the scale we are familiar with. On the atomic and subatomic scale the distinction between particles and waves becomes blurred, and an individual electron, for example, can be thought of as an extended pattern of waves, like ripples on a pond. It is this pattern of waves that surrounds the nucleus of an atom, and it is the properties of those waves that prevents an electron from 'falling in' all the way to the nucleus.

The full description of this strange behaviour of entities on the atomic scale is provided by the theory known as quantum physics, or quantum mechanics. This says that *every* entity – including a golf ball, and this book – has both a particle nature and a wave nature. But this particle–wave duality is not apparent on the everyday scale – the 'macroscopic' scale – for a reason that is clear from studies of the nature of electrons carried out since the 1920s, when quantum theory was first formulated.

The key point is that, although electrons can behave both as particles and as waves, there is a very precise relationship between the mass of a particle and its wavelength. Experiments can be set up to measure the wavelength of a 'beam' of electrons, and different kinds of experiment can be designed to measure the momentum of individual electrons in a beam. Momentum is the product of the mass and velocity of an individual particle, and is very much a particle property; measuring the momentum of an electron beam is like measuring how much thump a target receives from a stream of tiny cannon-balls. The curious result that emerges from quantum physics is that the product of momentum and wavelength is always the same, a number now known as Planck's constant, after one of the quantum pioneers, Max Planck. Planck's constant is very small by everyday standards, 6.626×10^{-27} erg seconds (this means that there are 26 zeros between the decimal point and the 6626 if the number is written out in full). But it constrains the behaviour of electrons precisely. If electrons in a beam are made to go twice as fast, for example, then the momentum of each electron is doubled; so, in response, its wavelength is halved, keeping the product (momentum × wavelength) the same. Because the mass of an electron is just over 9×10^{-28} of a gram, all the numbers that crop up in the wave–particle equations for electrons are roughly the same size, and both the wave and the particle aspects of the electron are important to its behaviour.

But the same wave–particle relationship, with the same Planck's constant, applies to everything. For electrons, which all have the same mass, we had to imagine increasing the momentum by increasing their speed; but the momentum (mass × velocity, remember) would also be bigger (and the wavelength correspondingly smaller) if we replaced the electrons by heavier particles. An object like this book has a mass of several hundred grams – many other everyday objects have masses of several kilograms – and because Planck's constant is so small the only way that the equations can be made to balance is by assigning a comparably small wavelength to the object. For all practical purposes, the wave equivalent of this book, or anything else in the macroscopic world, is so tiny that it can be ignored. This does not mean that a book, a person, a car or anything else does not have a wave nature, but that the particle aspect of its nature completely dominates, because Planck's constant is so small.

Of course, this raises the interesting question of *why* Planck's constant should be so small. The traditional answer is that it is just one of those things, a number (like other constants of nature) that is simply built in to the fabric of the Universe and that we have to accept as part of nature. Recently, though, many physicists have asked why the laws of nature should have the forms that they do, and whether the Universe had any 'choice' in assigning such numbers as Planck's constant the precise values that they have. As I shall discuss at length later, the possibility that the constants and laws of physics might not always have been as we now find them is being taken seriously, and has deep implications, both for the existence of life-forms like ourselves, and for the possibility that the entire Universe is alive. For now, though, what matters is that we already have enough information to see how an individual electron can surround an atomic nucleus.

Treating the electron as a little bubble of wave energy, it is easy to see how it spreads itself around a proton. Like the sound waves which resonate inside an organ pipe, or the vibrations of a plucked guitar string, the electron wave can set up a particularly stable standing vibration, or resonance. Other things being equal, a collection of particles will always settle into the most stable state available, and an electron wave will attach itself to an atomic nucleus in the form of a standing wave surrounding the nucleus. It is impossible to point to one place in that wave pattern and say 'the electron is there'; all we can do is say that 'the energy and charge of the electron are spread over that region of space around the nucleus'. The region of space around the nucleus where an individual electron spreads its standing wave is called a 'shell'. And it really is spread out. Roughly speaking, if the size of a nucleus is represented by a grain of sand, the electron cloud around the nucleus would be as big as the Albert Hall. Because the cloud of electrons largely screens the nucleus within from the outside world, the way in which atoms combine to form molecules – chemistry – depends almost entirely on the number of electrons and the distribution of their shells in the outer parts of atoms (but, of course, the number of electrons in an atom is the same as the number of protons in the nucleus).

Atoms of the same element can, in fact, come in slightly different varieties, called isotopes, even though they have identical chemistry. The simplest atom of hydrogen has just one proton in its nucleus, surrounded by one electron. But there is also a form of hydrogen, known as 'heavy hydrogen', or deuterium, which has a proton and a neutron stuck to each other in its nucleus, still surrounded by a single electron. Similarly, although every nucleus of helium contains two protons, and every atom of helium has two electrons surrounding the nucleus, some helium nuclei contain just one neutron, while others contain two. The neutrons are important because they help to hold the nucleus together, against the tendency of all the positive charge in the protons to blow it apart. Both protons and neutrons 'feel' a force of attraction known as the strong nuclear force, which makes them stick to one another. But it is easier for a proton to stick to a neutron, rather than to another proton, because of the electric force of repulsion between like charges. So there is no helium nucleus, for example, which contains only two protons, and apart from the simple single-proton hydrogen nucleus every nucleus contains neutrons as well as protons, and all but the lightest nuclei contain more neutrons than they do protons.

Outside the nucleus (any nucleus), there is room for two electrons in the closest possible shell, so both the electrons in a helium atom are effectively at the same distance from the nucleus. But for heavier elements, with more protons in their nuclei and more electrons surrounding them, things are more complicated. The next eight electrons can all fit in to one shell surrounding the nucleus and the inner two electrons, then any further electrons have to go into shells more distant from the nucleus. As far as life on Earth is concerned, however, the chemistry we are most interested in is the chemistry of carbon, and carbon has a nucleus containing six protons (plus associated neutrons) surrounded by six electrons, two in the innermost shell and four half-filling the next shell. So I shall ignore the interesting complications of the arrangements required to pack more than ten electrons around an individual nucleus.

The key to making molecules (combinations of atoms joined together) from particular atoms is the number of electrons in the outermost occupied shell of those atoms. For reasons I shall not elaborate upon here, but which may be fully explained within the framework of quantum mechanics (and outlined in my book *In Search of the Double Helix*), configurations in which the outermost shell is exactly filled with electrons are particularly stable, and atoms combine together as molecules in an attempt to reach this particularly stable state. For hydrogen, with just one electron in its outer shell, but room for two, one way in which this is achieved is by two atoms joining forces to make a molecule of hydrogen, written as H_2. Each nucleus gets a part share in two electrons, instead of a full share in one electron, and this creates the illusion that both places in the innermost electron shell are filled in each atom. Helium, on the other hand, is quite happy with its two electrons and has no chemical predisposition to alter the arrangement. This makes helium a very inert substance, chemically speaking, which scarcely reacts with anything. Helium gas is made up of solitary atoms (He), not molecules.

Carbon is special because it has six electrons, two tightly bound to the nucleus in the innermost shell and four in the second shell, where the ideal number would be eight. With a shell precisely half full, carbon has the greatest possible repertoire of chemical activity, with four electrons to share, so that one atom of carbon can combine with four other atoms. If there were only two or three electrons in the outer shell, then obviously the atoms could only make two or three connections with other atoms; slightly less obviously, if there were five or six electrons in the outer shell, the atom would still only make three or two connections, because that number of shared electrons would bring its quota up to the desirable eight.

Each connection between atoms is called a bond, and involves two electrons, one from each of the atoms linked by the bond;

every carbon atom can form four chemical bonds, and the simplest compound of carbon results when four separate hydrogen atoms attach themselves to one carbon atom in such a way that a pair of electrons is shared between each of the hydrogen nuclei and the carbon nucleus. Each hydrogen atom has the illusion of a filled inner shell of two electrons, while the carbon atom has the illusion of a filled second shell of eight electrons. Carbon atoms can also bond with each other, to form long chain molecules in which there is a 'spine' of carbon atoms, each held to its neighbours by pairs of shared electrons, with hydrogen atoms (or something else) stuck onto the sides of the chain. For an individual carbon atom in the chain, two of the four available bonds are used to hold onto the carbon atoms immediately ahead and behind in the chain, while two are free to join onto something else on either side. At the head and tail of the chain, of course, the end carbon atoms have three 'spare' bonds to use in this way. Such chains can even loop around on themselves to make a ring. The most stable ring configuration contains six carbon atoms 'holding hands' in this way, with other atoms joined onto the outside of the ring. But the most important carbon compounds for the story of life on Earth are the long chains, in which thousands of carbon atoms are linked together, with a variety of other atoms and molecules (including some of those ring compounds, and even other chains) sticking out from the sides. The chemistry of life is, essentially, the chemistry of carbon, and in particular the chemistry of long carbon chains with interesting bits and pieces attached to them.

The simplest working distinction between the living and the non-living is that living things are able to reproduce – they can make copies of themselves out of the raw materials they take in from their environment. The key life molecules today, which all of us carry in all our cells, are long chains, based on a carbon spine, with two strands twisted together to make a single molecule of deoxyribonucleic acid, or DNA. DNA has the neat ability to unravel itself, forming two separate mirror-image strands, each of which can gather up the necessary chemicals to build a counterpart strand, producing two DNA 'double helices' where there used to be one. This happens, for example, when a cell divides into two. As we shall see, DNA can also do other tricks that are essential for our kind of life, but this ability to replicate itself is the key to life. Life began when – somehow, somewhere – a combination of chemical reactions produced a molecule that was capable of making copies of itself by triggering further chemical reactions. From then on, the story of life has been one of competition between different life-forms for the available 'food' (the chemical elements and compounds necessary to make copies) and protection against other life molecules that might regard any complex compound (like the odd molecule of DNA floating around freely) as an easy source of raw materials for its own life processes. So the fundamental molecules of life are today concealed within a protective wall of material as individual cells. In some cases, such as bacteria, the cells themselves are each a complete living organism; in other cases, many millions of cells combine together to make up, say, a human being or a tree. How this came about is the subject of the next chapter – and it is all relatively straightforward once the first life molecules appear. Just how and where life originated, however, is still a subject of debate, although you might not think so from reading many school textbooks and popular accounts, which tend to present only one side of that debate.

The conventional theory of the origin of life goes back to the 1920s, to the suggestion by J. B. S. Haldane, and independently by the Russian Alexander Oparin, that the right carbon compounds (also known as 'organic' compounds, because of the importance of carbon chemistry to life) must have built up slowly in the Earth's oceans over a long period of geological time, until the complexity of the compounds being produced by chemical reactions grew to the point where the first life molecules (the first molecules capable of copying themselves) appeared. There may even, in this picture, have been a single 'first living molecule', from which all of life on Earth is descended. Indeed, the idea can be traced back still further, since although he never went public with his thoughts on the matter, in 1871 Charles Darwin wrote to a colleague, Joseph Hooker:

It is often said that all the conditions for the first production of a living organism are now present, which could ever have been present. But if (and oh! what a big if!) we could conceive in some warm little pond, with all sorts of ammonia and phosphoric salts, lights, heat, electricity, &c. present, that a protein compound was chemically formed ready to undergo still more complex changes, at the present day such matter would be instantly devoured or absorbed, which would not have been the case before living creatures had formed.

In Darwin's day, nobody yet knew of the importance of DNA as the life molecule; proteins are also complex molecules based upon carbon chains, and are fundamental to the functioning of living cells, although they cannot make copies of themselves. Allowing for the fact that Darwin knew nothing of the role of DNA, this brief paragraph is prescient on several counts, but the key aspect is that before there was any life on Earth there was, of course, nothing around to eat up any concentrations of the kind of organic material (such as protein) that may have been produced by chemical reactions. As Haldane put it, before life appeared the products of such chemical reactions could have accumulated uneaten until the primitive oceans (or Darwin's 'warm little pond') reached the consistency of hot dilute soup – a rather thin chicken bouillon, according to a more recent estimate by Leslie Orgel.

In the world today, if you leave a bowl of chicken soup lying around it will 'go bad' as it is attacked by micro-organisms and the rich resource of complex carbon compounds that it contains is used by those micro-organisms to make copies of themselves. But until the first 'living' molecule appeared, the primeval soup would have remained untouched; as soon as the first molecule capable of using the compounds in the soup to make copies of itself did appear, it would have had a vast amount of raw material – food – available, and could have made vast numbers of copies of itself before running into the problem of competition for a diminishing resource.

Many experiments have shown that a soup of this kind really could have been produced in the oceans and ponds of the young planet Earth under the conditions thought to have existed at the time. Most probably, the early atmosphere was produced by outgassing from volcanoes, and was largely made up of the kind of gases which are produced by volcanoes today. It would have been rich in compounds like methane and ammonia, plus nitrogen, carbon dioxide and water vapour, but with very little free oxygen, if any. Without any oxygen to form an ozone layer, energetic ultraviolet radiation from the Sun would have penetrated to the surface of the Earth, while violent electrical storms in the developing atmosphere would have provided energy in the form of lightning. Energy stimulates chemical reactions. Today, ultraviolet radiation is bad for living things, because it disrupts DNA (this is why it causes skin cancer in people, and one reason why there is concern about the thinning of the ozone layer caused by human activities). But when you start out with simple compounds, the stimulus of an input of energy can encourage the molecules to stick together to form more complex ones.

As early as 1929, Haldane had drawn attention to experiments in which ultraviolet radiation had been seen to encourage the build-up of organic compounds from a mixture of water, carbon dioxide and ammonia. In more carefully controlled experiments first carried out in the 1950s, and repeated with minor variations many times since, electrical discharges and ultraviolet light have been passed through sealed glass jars containing various mixtures of the kinds of gas thought to have dominated the early atmosphere of the Earth. Such experiments do not quite produce the proteins that Darwin speculated about (let alone DNA), but they do produce slightly simpler compounds, known as amino acids, which are also made of chains of carbon atoms and are themselves the building blocks of proteins.

One reason why building blocks of life can be made out of such simple initial chemical ingredients as carbon dioxide, water and ammonia is that those simple ingredients themselves contain the four varieties of atoms that dominate the structure of living things. The Archbishop of Canterbury, like every other living person, may be 65 per cent water (H_2O), but even the other third of him is mainly made up of just four kinds of atom, the same hydrogen (H) and oxygen (O) found in water, plus carbon (C) and nitrogen (N). It can hardly be a coincidence that carbon, nitrogen and oxygen are among the more abundant products of nucleosynthesis inside stars. Carbon will react with whatever is available, and if what is available is nitrogen and oxygen (plus the ubiquitous hydrogen) then any complex chemistry that is going on in the Universe, including the chemistry of life, is bound to involve the same four elements. A full 96 per cent of your body is made up of just these four kinds of atom (C, H, O and N, often abbreviated as CHON) in different combinations; the complexity of living molecules arises not from the fact that they contain a great variety of different kinds of atom (they do not), but from the fact that just these four kinds of atom can be combined in large numbers in very many different ways – thanks, in large measure, to carbon's proclivity for chemical bonding. So the combination of simple molecules with energy to encourage chemical reactions is indeed enough to produce molecules complex enough to be regarded as precursors to life, perhaps just one or two steps away from life itself, although no self-replicating molecule has yet been produced in any of these experiments.

But although such experiments show that a combination of simple compounds with energy may lead to the production of living molecules, they do not show that this is how life on Earth must have originated. And there is one particularly puzzling problem about this scenario. You would expect all of this activity – the formation of the Earth and its atmosphere, the filling up of oceans and ponds with water, the brewing of the primeval soup – to take a fair amount of time. Just how long is a matter of guesswork, but it can be put in perspective. Our planet formed, along with the Sun and the other members of the Solar System, about four and a half billion years ago. We know this with considerable certainty, from a variety of geological and astronomical evidence. The planet formed from an accumulation of rocky material, smashing together and sticking under the attraction of gravity to build up a larger body. At first the young Earth was pounded and kept sterile by the impact of more of these rocky asteroids, but eventually its orbit was swept clean and it began to cool and settle down into a stable state.

It is equally certain, from unambiguous fossil evidence, that life in the form of early bacteria was well established on Earth by four billion years ago, and there are indications that it had already been around for some time. This gives only, at most, a few hundred million years to do the trick of turning methane, ammonia and the rest not just into the first living molecule, but into actual living cells, in which all the important chemistry of life is protected from the outside world and tucked up safe inside the cell wall. Could life really have got going from scratch as quickly as that?

We may never know the answer for sure. But one of the most important developments in recent years in our understanding of the relationship between life and the Universe has been the realization that there may well have been no need for life to get going from scratch on the surface of planet Earth. The young planet may well have been seeded with, at the very least, much more complex molecules than methane and ammonia, if not with life itself.

I have already mentioned that studies of the spectrum of cyanogen, a compound of carbon and nitrogen found in interstellar clouds, could have revealed the presence of the cosmic microwave background radiation in the 1940s, if any of the astronomers who knew about the cyanogen spectrum and about Gamow's version of the hot Big Bang theory had put two and two together. The cyanogen is one of the constituents of vast clouds of gas and dust that lie between the stars in our Milky Way Galaxy, and other galaxies contain similar clouds of cool material. Right until the end of the 1960s, astronomers visualized these clouds as rather dull places, cold and dark and containing only very simple chemical compounds, including molecular hydrogen, cyanogen and the hydroxyl 'radical', a compound in which each molecule contains just one hydrogen atom and one oxygen atom (OH), like water (H_2O) with one hydrogen atom missing. These simple compounds are constantly bathed in energy from stars (in the form of starlight undimmed by any atmosphere or ozone layer) and other forms of radiation in space (cosmic rays). But just as nobody put two and two together in the 1940s to 'find' the cosmic microwave background, so nobody in the 1950s and 1960s put this knowledge together with the fact that experiments in laboratories here on Earth had shown that a combination of simple molecules with energy leads to the production of more complex molecules, to predict that there ought to be complex molecules in interstellar clouds. So it came as a surprise when a newcomer to the study of interstellar clouds suddenly started finding all kinds of complex molecules in space in 1968.

So that it may come as something less of a surprise to you, here is a brief reminder of where the material in those interstellar clouds came from. The COBE data show that very soon after the Big Bang itself the Universe contained vast clouds of hot gas already clumped into a hierarchical clustering pattern as a result of the physical processes which dominated the last stages of the radiation fireball. These clouds, containing a great deal of hydrogen, a smaller proportion of helium and hardly anything else at all, quickly settled down around concentrations of dark matter in the Universe, much of the matter in the clouds quickly collapsing to form stars. Many of these first-generation stars were very large, and ran through their life cycles very quickly, converting hydrogen and helium into heavier, more complex nuclei in their interiors by the process known as nuclear fusion. When these first stars ended their lives in violent explosions, the heavier elements were scattered among the interstellar material. Here new stars formed among the swirling gas clouds which, in the case of our own Milky Way and many other galaxies, settled into a spiral pattern, rather like the spiral pattern you get when you gently stir cream into a cup of coffee. I shall describe the processes by which stars are born and die, and the way in which the spiral pattern of stars in the Milky Way is maintained, more fully later in this book. What matters now is that the processes of star formation, nuclear synthesis, and stellar explosions have continued, albeit on a lesser scale, as the Galaxy has aged. So between the stars there are clouds rich in the elements produced by nucleosynthesis inside stars, including carbon, nitrogen and oxygen – as well as a great deal of unprocessed original hydrogen. These clouds are the stellar maternity wards: the places where new stars, and their attendant planets, form. Anything that is present in an interstellar cloud can become part of the input to a new solar system forming out of such a cloud, and all of the material on Earth and in your own body has been through this complex cycle of creation. In the words of the song 'Woodstock', it is literally true that we are made of 'stardust, billion-year-old carbon', and all the carbon in your body was processed inside stars not merely a billion years ago but well over 4.5 billion years ago, before our Solar System formed.

The person who opened the eyes of astronomers (and biologists) to the richness of the chemistry of interstellar clouds was Charles Townes, a physicist interested in microwaves, who had won a Nobel prize in 1964 as the co-inventor of a device known as a maser. A maser produces an intense beam of microwaves (the acronym stands for microwave amplification by stimulated emission of radiation), and the invention was the forerunner of a device that does the same thing with light, originally known as an optical maser, but now called a laser (from light amplification by stimulated emission of radiation). Townes was interested in radio astronomy at short radio wavelengths, of a few centimetres or so: the microwave region. Microwaves are widely used in communications links and radar, and it was radiation in this band which had, by the late 1960s, been identified as the echo of the Big Bang itself. But Townes was not particularly interested in the background of such radiation coming from all directions in space; rather, he wanted to probe interstellar clouds by doing the equivalent at microwave frequencies of taking the spectrum of the radiation from those clouds.

Radio waves are, like light (and, indeed, like X-rays and gamma-rays), part of the electromagnetic spectrum. The only difference between light waves and radio waves is that radio waves have a longer wavelength (which means a lower frequency, since frequency is just one over the wavelength). The wavelength is exactly what the name implies – think of radio waves as resembling a pattern of waves on the ocean, marching up towards the beach; the wavelength is the distance from the crest of one wave to the crest of the next. The processes whereby atoms and molecules absorb or emit radiation are the electromagnetic equivalent of the way in which a tuned guitar string resonates when it is plucked to produce a particular note, which represents a particular wavelength (or frequency) of sound waves in the air. Just as a shorter guitar string, held down at a certain fret by the finger of the guitarist, produces a higher-pitched note (that is, one with a shorter sound wavelength), so a smaller molecule resonates at a shorter wavelength of electromagnetic radiation. Individual atoms and very small molecules, indeed, are so small that they 'resonate' with light waves, while larger and more complex molecules resonate with short-wavelength radio waves, or in the infrared part of the spectrum.

In both cases, the resonance can either lead to the emission of electromagnetic radiation of a particular wavelength, as an atom or molecule gives up energy (like a plucked guitar string sounding a note), or it can lead to the absorption of electromagnetic radiation at a particular wavelength, as the atom or molecule absorbs energy (just as a loud note played through an amplifier may cause the strings of a nearby guitar, untouched by the player's hand, to vibrate in sympathy with the sound). With atoms, this leads to the characteristic 'fingerprint' spectrum described in Chapter Two, with a pattern of either bright or dark lines in the optical spectrum of rainbow colours. Large molecules produce the equivalent of 'lines' in the microwave spectrum – either peaks of radio noise at particular wavelengths, or gaps, wavelengths where relatively little noise is heard because radio energy from behind the interstellar clouds is being blocked by the resonance of molecules in the clouds. And all of this, of course, follows on very neatly from the work by Townes on microwave emission which had earned him his Nobel prize. He had just the experience needed to pull in the weak microwave signals from space and make sense of them – which was just as well, since the task was by no means easy.

There are two main problems in looking for complex molecules – 'polyatomic' molecules – by tuning radio telescopes to investigate such subtle details of microwave radiation from space. In order to study the spectrum of the radiation, it is, obviously, necessary to be able to tune your antenna and receiver system to a particular wavelength, and then to scan the microwave waveband to find the peaks and troughs in the signal that are the radio equivalent of spectral lines. The first problem is that, in order to detect a weak signal, the antenna has to be large; but in order to measure the intensity of the radiation at a precise wavelength, the antenna has to be very smooth – the shorter the wavelength, the smoother the antenna has to be. If you are looking for radiation with a wavelength of a centimetre or so, then the surface of the antenna has to be smooth, across its entire surface, to within a centimetre, or the incoming radiation will not be focused on to the receiver but will be scattered erratically by what appears to the radiation to be a rough surface. Optical telescopes have to be 'smooth' compared with the wavelength of light, substantially less than a millionth of a metre, and there is a limit to how big a telescope mirror can be built to this accuracy, which limits the light-gathering power of optical telescopes. Similarly, a very large radio telescope tends to sag under its own weight, making it difficult to maintain smoothness; but microwave radiation from space is weaker than the light from the visible stars, so the antenna has to be as big as possible in order to catch as much radiation as possible. A microwave antenna is a compromise between size and consistent smoothness. The receiver's amplifier system then has to boost the weak signal to observable levels without distorting the information it contains, and it was only in the 1960s that suitable receiver – antenna systems were developed for this work, building in large measure (like the discovery of the background radiation by Penzias and Wilson) on the expertise gained from the first attempts to broadcast signals around the world with the aid of communications satellites – and in Townes' case, on his long experience with microwave systems.

The second problem, which is still a problem today but was a much bigger problem in 1968, is that although the microwave radio 'lines' can identify a polyatomic molecule as precisely as a fingerprint can identify a human being, it is first necessary to 'take the fingerprint' by making the appropriate compound in the laboratory and studying its microwave spectrum at first hand. All the spectra of individual atoms have long since been analysed here on Earth, and the details are there in the textbooks, handy for optical astronomers to use. But the more atoms a molecule contains, the more subtly different electromagnetic resonances it is likely to be associated with, making the fingerprint distinctive but the effort needed to calibrate it tedious. It was not until some complex molecules had been discovered in space that the effort of identifying others seemed worth while – nobody wants to spend hours in the laboratory determining the microwave fingerprint of a molecule, then go and make the radio observations, only to find that the molecule is nowhere to be found in space. So when a Berkeley team inspired by Townes set off on the trail of polyatomic molecules in space in 1968, they started small. Their dramatic and almost instant success soon encouraged others, however, and the whole business took off in a big way in the 1970s as more and more molecules – almost a hundred, by the end of that decade – were identified in interstellar clouds.

First came ammonia, a relatively simple molecule containing just one atom of nitrogen and three of hydrogen (NH_3). This was identified in December 1968 by its radiation at a wavelength of 1.26 cm; it is hardly really a polyatomic molecule, but it showed that the system worked, and the molecule had never been identified in space before. It is also interesting as an established precursor of life, an essential ingredient in the manufacture of amino acids. Then the same team found the microwave fingerprint of water (H_2O) in space – still not really a surprise. The cat was firmly put among the astronomical pigeons, however, early in 1969, when another team found a pattern of radio 'lines' in microwave radiation from space which exactly fitted that of the molecule formaldehyde, one of the few that had already been studied in this way on Earth at that time.

Formaldehyde is not just a complex molecule by the standards of molecules known to be in interstellar clouds at that time. It is also an organic compound (which means that it contains carbon; its chemical formula is H_2CO), and was already specifically known to be one of the molecules that can easily be made by adding energy to a mixture of simple compounds in laboratory experiments. At this point the penny dropped, at least as far as some astronomers were concerned, and 1969 marked the beginning of a new sub-branch of astronomy, 'astrochemistry'. Simple compounds in interstellar clouds plus energy could indeed, they realized, make the precursors of life as efficiently in interstellar clouds as they could here on Earth. The bonus is that, whereas the entire history of the Earth covers no more than 4500 million years, of which less than 500 million had been years available for the build-up of complex chemicals before life appeared, a cloud in space might have been around for thousands of millions of years before the Earth even came into existence. It could do the chemical mixing job required before life itself came on the scene, and greatly shorten the interval needed for life to develop here on Earth if, as seems likely, the Earth were 'seeded' by complex chemicals early in its history.

One possibility is that complex chemicals from space may have been brought down to Earth, soon after the planet formed, by the impact of comets. Comets are rather like huge dirty snowballs, great lumps of icy material, which occasionally plunge into the inner regions of the Solar System and swing past the Sun before returning again to the depths of space. Astronomers calculate that there must be a great shell of cometary material, known as the

Oort Cloud, surrounding the Solar System, out beyond the orbit of the outermost planet, and that occasionally one of these interplanetary icebergs is dislodged from its orbit by an interaction with another body (perhaps by gravitational perturbation by a nearby star), and is steered inwards to make its dive past the Sun.

Sometimes, a comet diving into the inner part of the Solar System collides with the Earth – this is possibly what caused the huge explosion over Tunguska, in Siberia, on 30 June 1908. Comets contain, frozen in cold storage, the almost unsullied raw material of the interstellar medium, left over from the clouds of material from which the Solar System formed, and there must have been many more of them around when the Solar System was forming, with quite frequent collisions between the young planets of the Solar System and comets dislodged from the Oort cloud. Any complex molecules found in interstellar clouds are likely to be present in comets, and therefore likely to have arrived on Earth when these icy bodies impacted with our planet soon after it formed. Even quite delicate complex molecules could survive the journey through the Earth's atmosphere and impact with the surface if they were locked away inside chunks of ice, which would act like the heat shield on a spacecraft re-entering the Earth's atmosphere, protecting the delicate astronauts. Another possibility is that fluffy grains of material from the tail of a comet might float gently down through the atmosphere, like thistledown. Either way, the young Earth – or any suitable young planet – would be enriched with organic material.

Christopher Chyba and colleagues from Cornell and Yale Universities have estimated that soon after the Earth formed the amount of organic matter raining down on it from space may have been as much as 10,000 tonnes per year; in just three hundred thousand years this would add up to as much mass as all the living things on Earth today put together. Edward Anders of the University of Chicago puts the figures in a slightly different context – in the few hundred million years between the last cataclysmic asteroid impact associated with the formation of the Earth and the date corresponding to the first traces of life in the fossil record, he says, enough organic cometary material would have reached the surface to cover the Earth with a layer containing 20 grams of organic material for every square centimetre. Whichever way you look at it, the precursors of life would be present even in the earliest primeval soup, and it would not be long, compared with the age of our planet now, before self-replicating molecules appeared. The big outstanding question, though, is just how complex the molecules that seeded our planet from space may have been.

Dozens of polyatomic molecules have now been identified in space. Almost all of them are carbon compounds, in line with the known chemistry of carbon and its prolific ability to combine in different ways with many other elements (and with itself) that makes it so important to life. Some of these molecules, including formaldehyde, have now been detected even in the microwave radiation reaching us from other galaxies, and there is every reason to believe that carbon-based complexity is a common feature, not just of the Earth or even our Galaxy, but of the Universe at large. If carbon-based life-forms such as ourselves can evolve in our Milky Way Galaxy, then there is no reason why carbon-based life-forms like ourselves could not have evolved in, for example, the galaxy known as NGC 253, in which our radio telescopes have identified clouds containing formaldehyde. We may never know for sure that there is other intelligent life in the Universe, but the odds now seem to favour it since these discoveries indicate that the same chemical processes, right up to the complex chemistry of the precursors to life itself, go on throughout our Galaxy and in other galaxies as well. It is a curious (and comforting, to my mind) thought that somewhere in the galaxy NGC 253 there may be carbon-based life-forms with DNA, or something very similar, in their cells, studying the heavens with radio telescopes, and that they may be speculating on the significance of the detection of formaldehyde emission lines from the rather unspectacular collection of stars which it pleases us to call 'our' Galaxy.

At the very least, the young Earth must have been laced with molecules such as H_2CNH, HCCCN, H_3CCOH and others. Furthermore, some of the molecules already identified in space are known to combine easily with one another to make yet more complicated molecules. For example, formic acid (HCOOH) and methanimine (H_2CHN), both identified in dense clouds in space, react to produce the simplest amino acid, glycine (NH_2CH_2COOH), which is only one step away from being a life molecule. Formaldehyde, remember, was the key discovery in all this excitement, because it is an organic molecule. It is itself a common component of bigger and more complex organic molecules, including sugars, which are essential to life as we know it. Quite probably, even amino acids were present in the primeval soup, virtually from the moment the Earth cooled to the point where puddles of liquid water could lie on its surface.

All this has prompted Jim Lovelock to comment, in his book *Gaia: A New Look at Life on Earth*, that 'it seems almost as if our Galaxy were a giant warehouse containing the spare parts needed for life'. Given a planet as rich in the components of life as it now seems the early Earth must have been, chemical reactions could easily have produced a replicator molecule within 500 million years. The odds against the first replicator molecule appearing may be astronomical, but it is now clear that the time available was also astronomical, with some ten billion years of 'prehistory' of chemical conditioning inside interstellar clouds possible before those warehouse components ended up in the warm little ponds on the surface of our planet. As Lovelock puts it, 'life was thus an almost utterly improbable event with almost infinite opportunities of happening', so it happened, eventually.

But these dramatic conclusions about the origin of life on Earth and its links with the Universe at large seem to have passed almost without comment into the pages of science books and journals. This is astonishing. More than twenty years after the discovery of formaldehyde in space, most scientists, let alone most people, seem unaware of the literally vital importance of the discovery, of how it solves the last great riddle in our understanding of the origin of life (the speed with which life emerged on Earth after the planet formed), and of how it links the living Earth to the living cosmos. What is doubly astonishing is that hardly anybody seems to be aware that this breathtaking and well-founded suggestion that the young Earth was seeded with organic molecules, probably as complex as amino acids, is itself now the most *conservative* interpretation of the evidence that complex chemical processes have been going on in space for billions of years longer than the Earth has existed. The notion that life had to start from scratch (which means from methane, ammonia, carbon dioxide and energy) in warm little ponds on the surface of the Earth is now completely outmoded, and those opponents of evolutionary ideas who still, even in this scientific age, argue that life has not had time to evolve on Earth are, along with those misguided scientists who try to defend the position, simply wasting their breath. It is scarcely necessary to go beyond the probable arrival of amino acids and similarly complex molecules in the prebiotic soup in our search for how life got started on Earth. Two alternatives, however, have been put forward, each by maverick scientists who have made their distinguished mark in other ways, in which not just the precursors to life but living cells themselves might have seeded the Earth from space.

The first possibility is that genuinely living forms – replicators – may have become established in interstellar clouds during the first ten billion years of the history of the Universe. Life, in this picture, originated in space, is still common in interstellar clouds, and only moved down to seed planets once suitable planets appeared in the evolving Universe.

The implications are profound. If the kind of replicators we are descended from started replicating in interstellar clouds, then it is all the more likely that life everywhere is descended from the same sort of replicators. This argument was put forward forcefully by Fred Hoyle (the same Fred Hoyle who explained how the elements are made inside stars) and his colleague Chandra Wickramasinghe in the 1970s. Unfortunately, though, Hoyle and Wickramasinghe shot themselves in the foot with their claim not just that life may have originated in interstellar space, but that diseases such as influenza and bubonic plague may have reached the Earth from comets, wreaking havoc upon a population that had never previously encountered the organisms that caused those diseases. This caused a furore among medical experts who, angry with two astronomers who were trampling all over someone else's territory, soon established beyond reasonable doubt that such diseases have a terrestrial origin, and cannot be blamed on viruses spread by comets. Unfortunately, this typecast the whole package of ideas presented by Hoyle and Wickramasinghe as distinctly on the cranky side, if not completely loony.

It is even possible that the dust stirred by the argument may have cost Hoyle a Nobel prize. The discovery of the way elements are synthesized in stars was undoubtedly one of the greatest achievements in astrophysics, and it is equally beyond doubt that Hoyle was the leader of the team that made the discovery. Yet in 1983, before the dust from the 'diseases from space' storm had settled, the Nobel Prize for Physics was awarded for this work to Hoyle's long-time colleague Willy Fowler. It is impossible to conceive of any reason why the Nobel committee should have snubbed Hoyle in this way, except for their fear that if they had given the prize jointly to Fowler and Hoyle it might have provided a platform for Hoyle's latest wild notion, and lent credence to his hypothesis that plagues come from comets.

This was a double injustice. First, because no matter how cranky Hoyle's later work may (or may not) have been, the nucleosynthesis work still stands as a beautiful piece of physics, well worthy of a Nobel prize. Secondly, like everybody else then and since, the Nobel committee seems to have missed a crucial point. Hoyle and Wickramasinghe do indeed seem to have erred in pushing their ideas about life and the Universe too far with the notion that diseases strike from space and that evolution on Earth may have been influenced by the continual arrival of new kinds of living material carried by comets. But their underlying contention that life itself may have first appeared in interstellar space, and may have evolved there to the stage of single-celled bacteria, is soundly argued, based on good scientific evidence, and may very well be correct. I shall say no more about the crankier aspects of the Hoyle–Wickramasinghe argument. But I am convinced that, although their theory of the origin of life is still new, and may have to be altered (perhaps drastically) as new discoveries are made (and when other astronomers and evolutionary biologists have the courage to take the ideas seriously and do some real work along the lines sketched out by these two pioneers), it is the best available answer to the question of where the first replicator came from, and well worth spelling out for you here. The theory starts out from the evidence that there are not only clouds of gas between the stars, but vast amounts of dust as well.

These dust patches literally block out the light from stars behind them, so that often the densest patches appear as black pits against the surrounding, star-dominated sky; but even where there is no patch of dust along the line of sight dense enough to block out the light entirely, hardly any starlight reaches us without being affected by more tenuous regions of dust which it encounters on its way here. As well as being dimmed a little by the ubiquitous dust in space, the colour of starlight is affected. Short wavelengths of light – at the blue end of the spectrum – are scattered more easily by the dust, while longer wavelengths – at the red end of the spectrum – can get through relatively undisturbed. So dust in space literally reddens the light from stars behind it. This is quite different from the redshift effect, which bodily shifts features in the spectrum from the blue end towards the red end. Indeed, with those far-distant objects known as quasars, which actually emit a lot of energy in the ultraviolet part of the spectrum, at even shorter wavelengths than blue light, one consequence of the redshift is that the images of these objects look very blue in our telescopes, because the spectra that we see have all of this energy redshifted from the invisible ultraviolet part of the spectrum to the visible blue part.

By contrast, the reddening of starlight by interstellar dust is exactly the same process as the reddening of a sunset by dust in the atmosphere of the Earth, which also scatters the blue light and lets the red light pass almost undisturbed. And it is because the shorter-wavelength blue light is more easily scattered than the longer-wavelength red light that the sky is blue – the blue light from the sky is simply sunlight that has been scattered by the Earth's atmosphere so much that it comes to us from all directions in the sky (something that was proved, incidentally, in the first decade of the twentieth century, by a young physicist called Albert Einstein).

The amount by which starlight is dimmed and reddened overall on its way to us reveals how much dust there is in the intervening regions of interstellar space, and as usual the amount of dimming at particular wavelengths can be used to yield information about the kinds of particle that are doing the dimming. It has been known since the 1930s that the dust particles responsible are tiny grains, about the size of the wavelength of visible light, which is measured in nanometers (nm); one nanometre is a billionth of a metre, or a millionth of a millimetre. Visible light covers a band of wavelengths spanning a few hundred nanometres, so the size of an individual grain of interstellar dust is around one hundred-thousandth of a centimetre.

Even so, there is a vast amount of this interstellar material in our Galaxy. About 2 per cent of all the mass of interstellar clouds is dust, with the rest mainly hydrogen and helium gas, forming the great clouds which are the birthplaces of stars and solar systems. The dust in interstellar material is a product of the nucleosynthesis in earlier generations of stars, which we know (from the work inspired by Hoyle in the 1950s) produces carbon, oxygen and nitrogen in relative profusion, so the dust grains in interstellar clouds must be made, in large measure, of combinations of these three elements with hydrogen (the helium, remember, is very stable, 'needs' no extra electrons, and scarcely reacts with anything). Including all the gas, interstellar matter makes up about one-tenth as much mass as all the bright stars in the Milky Way put together. This adds up to ten billion (10^{10}) times the mass of our Sun, and even 2 per cent of this is still an impressive 200 million solar masses of material spread around our Galaxy in the form of tiny dust particles in interstellar clouds.

Since the 1930s, observations of the amount of dimming produced at different wavelengths – the amount and nature of the reddening of starlight by interstellar dust – have been extended and improved. Today we even have measurements of the absorption by interstellar dust at wavelengths beyond the visible spectrum, in the infrared and ultraviolet. These measurements are made by instruments carried on rockets and satellites above the Earth's atmosphere, which obscures radiation at these wavelengths. They show that the strongest absorption is at around 220 nm, in the ultraviolet. Before this discovery was made in the mid-1960s, the most widely favoured explanation of interstellar grains saw them as icy particles, a kind of 'snow' of frozen water, methane and ammonia. But none of these three candidates blocks radiation particularly strongly at this wavelength, and none of the spectral lines that really are associated with such icy particles have actually been found in the reddened light from the stars.

Hoyle and Wickramasinghe argue that it is, in any case, very difficult to make ice grains or snow in space, and that it would be a lot easier to get dust clouds made up of tiny particles of carbon – in effect, soot. This may seem bizarre, but carbon is one of the major products of stellar nucleosynthesis. There is a family of stars called carbon stars which are shown by their spectra to have atmospheres rich in carbon atoms, and which vary regularly in brightness, with periodic fluctuations of a year or so. The natural explanation is that many (if not all) stars go through such a phase of activity, with only a few visible as carbon stars at any one time. All stars are constantly losing matter from their atmospheres, and the periodic fluctuations of carbon stars are interpreted as a sign that they are 'breathing' in and out, puffing a stellar wind of material, rich in carbon, out in to space. The idea is now strongly supported by the discovery of spectral features corresponding to graphite (carbon particles, like the inside of an ordinary pencil, which contain many atoms joined together) in the radiation from carbon stars.

In the early 1990s, astronomers found evidence of an even more complex form of carbon in interstellar dust. A previously unexplained pattern of spectral lines was interpreted by researchers from the NASA Ames Research Center as coming from molecules made of sets of a dozen atoms of carbon joined together with ten hydrogen atoms to make a structure known as pyrene ($C_{12}H_{10}$). In the laboratory, the NASA team measured the spectrum of pyrenes associated with substances such as argon and neon, and found that the shape of the infrared spectrum closely matched previously unidentified features of the infrared spectrum from interstellar material. Similarly complex molecules are needed to explain other features of this interstellar absorption of infrared radiation.

There is no doubt at all that interstellar dust contains not just carbon atoms but more complex structures in which many atoms are linked together. Some of these molecules may be in the form of long chains, rather than rings. Carbon also explains the strong interstellar absorption at 220 nm – so well, indeed, that some astronomers argue that *only* carbon grains are needed to produce the interstellar reddening. But there must be nitrogen and oxygen present in the clouds too, quite apart from the hydrogen gas which is the dominant component. There ought to be molecules composed of all these kinds of atom present in the clouds to some degree, and Hoyle and Wickramasinghe have shown just which molecules can best explain the observations, and how they may have formed in interstellar space.

At this point, the story switches to studies of the infrared radiation from clouds of gas and dust which are actually collapsing to form new stars, and to the 'cocoons' of dusty material surrounding young stars. These clouds are heated by the energy coming out from the forming, or newly formed, stars inside, and reach temperatures of several hundred degrees Kelvin, providing energy which could stimulate chemical reactions. Superimposed on the spectrum of the radiation from these clouds are emission features corresponding to radiation from the individual particles in the dusty clouds. Three strong infrared features dominate the spectra from many of the hot clouds of dusty material associated with the birth of stars. They occur at 2–4 μm (micrometres), 8–12 μm and around 18 μm (a micrometre is a billionth of a metre). Various attempts have been made to explain these features, usually by suggesting that each of the three features is produced by a different set of chemical compounds in the clouds. But Hoyle and Wickramasinghe found that one single compound can explain the entire spectrum. What just about every other astronomer found hard to swallow, though, is that this single compound whose radiation can explain the entire infrared spectrum of the hot dust around young stars is cellulose – a fundamental part of the structure of plants here on Earth.

But although cellulose itself is very much a biological molecule, it is a member of a large family of molecules known as polysaccharides, and it is possible that polysaccharides very similar to cellulose could be produced without invoking life processes. Since the structure of all polysaccharides is much the same, they have very similar infrared spectra. Although Hoyle and Wickramasinghe hit the scientific headlines by suggesting that cellulose itself could explain the pattern of radiation from the dust around young stars (almost seeming to imply that those dust clouds contain living plants), the slightly less dramatic (and considerably more plausible) implication is that some other members of the same polysaccharide family are actually responsible.

The basic component of a polysaccharide chain is a so-called pyran ring, a hexagon made up not of six carbon atoms 'holding hands', but of five carbon atoms and one oxygen atom (C_5O). These rings easily join together in a chain, with one of the carbon atoms linking to another oxygen atom, which itself also binds to a carbon atom in the next pyran ring in the chain. And although it is a tedious process to make one such chain by sticking atoms together, once formed a pyran ring shows one of the fundamental qualities of life – it acts as a pattern, or template, which encourages the formation of more rings, which link up in a growing chain. When the chain splits into two or more pieces, each piece then continues growing (as long as the necessary chemical ingredients are available), pulling carbon and oxygen out of the surrounding chemical mixture to make more pyran rings.

Now, although nobody would argue that a simple pyran ring, or a polysaccharide chain built up of pyran rings, is alive (its behaviour is more like that of a growing crystal), this behaviour ensures that under the right conditions a great deal of the carbon and oxygen available is turned into polysaccharides, just as under the right conditions a great deal of the carbon, nitrogen, hydrogen and oxygen on Earth has been turned into living cells. Could the right conditions include the environment of hot cocoons of dust around young stars?

If so, where is the nitrogen that also ought to be present in the interstellar clouds if our understanding of stellar nucleosynthesis is correct? Again, spectral studies hold the key. A broad spectral feature at about 443 nm, stretching over a width of 3 nm, can be explained in terms of the presence of a complex molecule built around four five-sided rings, each made up of four carbon atoms and one nitrogen atom (C_4N). The wealth of prebiotic material which could be present in interstellar clouds is astonishing, and, throwing caution to the wind, Hoyle and Wickramasinghe have suggested that genuine life-forms may have evolved in cometary clouds even before the precursors of life were brought to Earth by collisions with comets.

Although conditions in the cometary cloud are cold enough to freeze water unless some source of heat is available within the comets themselves, when large stars explode as supernovae and seed the interstellar medium with heavy elements there is a scattering of radioactive material among the debris. Radioactive atoms (or isotopes) differ from stable atoms because they are formed with an excess of energy, which they give up as they convert themselves into stable isotopes, a process known as radioactive decay. If comets are made of material ultimately derived from a supernova event, as seems likely, they would initially contain radioactive elements from the preceding supernova (or supernovae), and in particular the isotope aluminium-26. The radioactive half-life of aluminium-26 is 700,000 years, which means that it takes that long for half the original atoms to give up their surplus energy and turn into stable isotopes. The heat produced as the radioactive aluminium decayed in this way could have warmed the hearts of cometary nuclei, melting water to provide exactly the warm little ponds that the piecing together of the first living molecules may have required – as many as 100 billion separate warm little ponds, with 100 billion separate chances for life to develop, even if we restrict ourselves to the Oort Cloud around our own Solar System. And if life emerged somewhere in space, in a cometary pond or by some other process that we cannot yet even guess at, then the constant stirring of interstellar material by stellar explosions and by the passage of stars and planetary systems through the clouds of gas and dust would have ensured that it reached every suitable planet.

Even if you cannot accept this idea of an interstellar origin of life, it still does not necessarily follow that life evolved out of non-living material here on Earth. Assuming that it is indeed possible for non-living material to evolve into living material (which our existence suggests to be the case), why could this not have happened first on some other planet in the Milky Way Galaxy long before our Solar System formed? Francis Crick (who shared a Nobel prize for his discovery, jointly with James Watson, of the now-famous double helix structure of the life molecule, DNA) turned his attention to this possibility in the 1970s, first in a scientific paper with his colleague Leslie Orgel, and then in a popular book, *Life Itself*. His contention is that by some 10 billion years after the Big Bang, 4 billion years ago, at around the time our Solar System was forming, there already existed at least one civilization in our Galaxy more advanced than we are today. Perhaps by sending unmanned space probes to other planets, they may have learned that while many potential homes for life existed in the Galaxy, they were all uninhabited. And, motivated by a desire to ensure the continuity of life even if their own civilization were to collapse, they may have seeded those planets with life.

The best way to achieve this, according to Crick, is to send bacteria – just the kind of bacteria known to have been around on Earth almost as soon as the planet cooled – to be dumped on young planets from automated space probes. Such bacteria can survive the freezing cold of the journey through space, and can thrive on a young planet without oxygen; they would carry the DNA code of life to new worlds, where they could proliferate and evolution could be left to do its work.

Now, this really is a way-out idea at the very fringe of scientific respectability. It sounds like science fiction, yet in a few decades we shall ourselves be in a position, if we wished to do so to send rocket-loads of simple bacteria to be dumped on promising young planets out there in the Milky Way. Just as Hoyle, an astronomer, annoyed the biologists by encroaching on their territory, so Crick, a biologist, has annoyed the astronomers with *his* suggestion. But Crick's ideas put those of Hoyle and Wickramasinghe in a new light. The possibility that life first appeared in space seems rather less extreme, and correspondingly more plausible, than the notion that our planet was deliberately seeded from space. It cuts out the 'middle man' of planetary life entirely, and means that the chemistry leading up to life could have got to work in clouds in space within a billion years of the Big Bang (at around the epoch of the ripples in time revealed by COBE), giving ample time for the required complexity to emerge. And both these somewhat wild-eyed ideas put the third alternative in its proper perspective as a sober and serious contender. It is impossible, now, to see how the interstellar material from which the Earth and the rest of the Solar System formed could *not* have contained compounds a lot more complex, and interesting, than simple ammonia and methane, and it is certain that the interesting compounds that do exist in space were brought to Earth by cometary impacts. And once again, that extends the timescale of evolution from a mere 4 billion years to a full 14 billion years.

All the ideas concerning the origins of the first replicators are, to some extent, speculative; but all now embrace the Universe at large, and only the most blinkered Earth-chauvinist would try to argue that life got started from scratch on our own planet. The modern understanding of the evolution of the Universe, stars and planets is not at all embarrassed by the existence of life, but can explain in a general way how replicators came into existence, wherever their original warm little pond was located. It is no surprise to modern astronomy to find life like us on a planet like the Earth. And while the lack of direct evidence may make it impossible ever to determine exactly how life first got a grip on Earth, we now have a very clear idea of how life developed – evolved – from the state for which we do have the earliest direct evidence up to the present day. This covers more than 3 billion years of the Earth's history (perhaps 20 per cent of the history of the Universe so far), and takes in the development of life on Earth from single-celled organisms to ourselves.

CHAPTER FOUR

Life on Earth

Where do we come from? The complexity and variety of life on Earth today results from the natural processes of selection operating among replicators – some replicators are more efficient at reproducing than others, and some survive better than others. Over the course of three billion years, natural selection has led to the diversification of species and to the production of multicellular organisms. But some 'old' biological systems continue to reproduce effectively in their own ecological niches, and are not replaced. Alongside the 'modern' multicellular organisms like ourselves, we can still find types of single-celled species that are descended almost unchanged from the first colonists of the Earth.

The point is that nothing *wants* to evolve. The basic process of life is replication – copying existing molecules as accurately as possible. The success of replication at this level is shown by the continued presence on Earth of the 'living fossil' single-celled species that have remained essentially unchanged since the beginning of the story of life on Earth. Changes only happen by *mistake*, through copying errors; and very few of the changes are beneficial. By and large, imperfect copies of replicators are less successful than the originals, and do not survive, but instead end up as chemical food for successful replicators. Very occasionally, though, an imperfect copy turns out, entirely by chance, to be better than its original at the job of converting chemical food into replicas of itself. Such rare beneficial mutations will not only survive, but will spread through the environment.

Over many millions of years, the accumulation of such rare beneficial copying errors gives rise to species as diverse as a fly and a pine tree. But the process is arbitrary, and in the replicating business the single-celled species that have reproduced in the same form for billions of years could well be regarded as more successful than the collection of bizarre mistakes that has produced you and me. In both cases, though, what matters is that the basic process is one of mindless copying, with occasional mistakes – variations on the original theme – and then selection, by purely natural and equally mindless processes, of beneficial mistakes over harmful mistakes. Wherever this kind of copying and selection occurs, evolution can do its work. As we shall see later, this may mean a dramatic reappraisal of our instinctive notions of what is alive and what is not alive. But for now, I want to put the evolutionary process into perspective by looking in more detail at how it works for life on Earth.

To understand just how the replication process produces the mistakes that lead to evolution, we need to know a little more about the life molecule itself, deoxyribonucleic acid, or DNA. DNA is the basic copying material for essentially all life on Earth. A bacterium, a fungus, an ear of corn and a human being are all built according to specifications laid down in molecules of DNA within their cells. Simple, single-celled organisms reproduce simply by copying all the DNA in the cell and then splitting the cell into two daughter cells, each containing one copy of the genetic material. For multicellular organisms such as ourselves, things are a little more complicated, and reproduction involves special cells which separate off, carrying the genetic information with them in the form of copies of the DNA from the parent.

Within the cells of all but the simplest organisms, the DNA molecules are arranged in larger units called chromosomes. Each chromosome is a separate entity, like a separate book in a library. Specific bits of DNA within the chromosomes are called genes. They are like individual chapters of the book, helping to provide the detailed information on, say, whether an individual has blue eyes or brown, dark skin or fair.

Every DNA molecule is constructed in the famous double

helix structure. Two long chains, each built up from a carbon spine, twist around each other, and pairs of molecules link up from one helix to the other, producing a structure very like a spiral staircase in which the cross-bonds are the steps. Each individual DNA strand in the double helix is made up of many copies of just four chemical building blocks, carbon-based sub-units that join together like popper beads. The four basic units of the DNA chain are collectively known as nucleotides, and their chemical names are adrenine, cytosine, guanine and thymine, but they are usually referred to, even by biologists, simply by their initials: A, C, G and T. You can think of them as different coloured beads strung together to make different patterns in a very long necklace, or as four different letters spelling out different words in a rather basic alphabet. The four chemical blocks occur in different ordering along the length of different genes and chromosomes, exactly as if the plans for building and maintaining the whole living organism were written out in a four-letter alphabet.

This four-letter alphabet provides the basis for the common DNA language of all living organisms on Earth. They all share the same four-letter alphabet and DNA language, which is convincing evidence that we are indeed all descended from one uniquely successful ancestor, whether that ancestor first appeared inside a comet, or in Haldane's primeval soup, or in Darwin's warm little pond. It is essentially true that the chemical processes that maintain the growth and well-being of, say, a turnip plant (by reading the DNA code in turnip chromosomes and acting upon the messages it carries) could read the information in your own genes, if those genes were inserted into the cells of the turnip plant in the right way. This is the basis of recent highly successful attempts at genetic engineering in which genes are transferred from one species to another. A disease such as diabetes can now be treated, for example, by inserting copies of bits of human genetic material into the genes of animals, so that the animals will produce the human insulin that diabetics need in order to survive.

You might think that a four-letter alphabet would be rather restrictive for writing out all the instructions needed to keep a human body operating, let alone the instructions needed to build that body from a single cell in the first place. But remember that computers are based on the even simpler language of binary arithmetic, a language which has just a two-letter alphabet, like Morse code. Anything written in our familiar human languages can, in principle, be transcribed into Morse code, including Shakespeare, the Koran, and the songs of Lennon and McCartney. And we all know how successful computers are at storing and processing information, even though they also work on a two-letter alphabet, a language in which the only answers to any question are 'yes' or 'no', represented by switches that can only be either open or closed.

So how much information does a single human cell contain, locked away in DNA molecules and written in a four-letter code? If the genetic code of a chromosome were, in fact, a binary code, then the number of 'bits' of information (yes/no answers) in a chromosome would be just twice the number of nucleotide pairs matched up across the spiral staircase structure (twice the number of steps on the staircase). With a four-letter alphabet, the number of bits of information that can be packed in is four times the number of nucleotide pairs. A single chromosome may contain five billion nucleotide pairs, and each human cell contains forty-six chromosomes (the *same* forty-six chromosomes in each cell, apart from the special case of cells used in reproduction). So how much information, in terms of books written in the roman alphabet, does a single chromosome's 20 billion bits represent?

The neatest way to get an idea of what is involved is to imagine a kind of guessing game in which you have to identify a particular character from the twenty-six letter Roman alphabet, plus the ten numbers from 0 to 9, by asking a series of questions with yes/no answers. If you represent the answer 'yes' by a '0' and 'no' by a '1', you are in effect working in the binary language used by computers.

Suppose that the letter you are looking for is J. The series of questions and answers might go like this:

1. Is the character a letter? Yes: 0.
2. Is it in the first half of the alphabet? Yes: 0.
3. Is it one of the first seven letters in the alphabet? No: 1.
4. Of the remaining letters (H, I, J, K, L, M), is it one of the first three? Yes: 0.
5. Of the remaining three letters (H, I, J), is it H? No: 1.
6. Is it I? No: 1.

So only the letter J is left after just six questions and answers. This means that in binary code we can represent the letter J by the string 001011. Six 'bits' of information specify one letter of the alphabet, which means that with 20 billion bits of information in one chromosome, the string of DNA nucleotides on one chromosome contains as much information as something over 3 billion letters of the alphabet. Printers will tell you that in a book like the one you are now reading there are, on average, about six letters in every word; so one chromosome contains the equivalent of 500 million words – the equivalent of 5000 books like this one each 400 pages long and with 300 words on a page.

So it actually makes more sense to think of even an individual chromosome as a library, rather than a single book. Forty-six chromosome libraries like this are, it seems, what you need to describe the construction, care and maintenance of a human body. Single-celled bacteria need to store less information and have smaller DNA libraries. There is less chance of a copying mistake occurring when they reproduce, and they are slow to evolve. But the more DNA there is in the cells, the more chance there is of a copying error occuring when cells reproduce. Copying errors are the stuff on which evolution by natural selection operates, and once creatures with long chromosomal DNA molecules appeared on Earth, the pace of evolution speeded up.

But what is evolution? Charles Darwin was set thinking along the lines that were to lead to his theory partly by reading, in 1838, the famous *Essay on the Principle of Population*, written by

Thomas Malthus and first published in 1798. Malthus pointed out that populations of breeding animals (including people) could in principle reproduce in what is known as a geometric, or exponential, way, with the population doubling at regular intervals. With human beings, for example, all this would require would be for each couple to produce four children, all of which survive and reproduce in their turn. Those four children would leave eight grandchildren of the original couple, sixteen great-grandchildren, and so on. In this way, even a pair of the slowest-breeding animal on Earth today, the elephant, could produce a population of 19 million descendants in only 750 years, starting from a single, isolated breeding pair. And yet, until humans came along and upset the balance that had persisted for millennia, the number of elephants stayed much the same from century to century. On average, each pair of elephants alive 750 years before had left just one pair of descendants in Darwin's day. Why?

Clearly, the reason is that very many members of a population die without reproducing. In the natural state, survival to reproduce is the exception, not the rule. Darwin wondered why some particular individuals (the minority) should survive and reproduce – what distinguished them from their unsuccessful cousins? He saw that the individuals that survived and reproduced must be the ones best fitted to their environment (this is the origin of the famous phrase 'survival of the fittest'). Those that were less well adapted would lose out in the competition to get food, or a mate, or a place to make a home – or they might simply get eaten by predators.

The keys to Darwin's theory of evolution by natural selection are that there are indeed a large number of individuals in each generation, that those individuals show a variety of characteristics (a variety of survival skills, if you like) which make them more or less suited to their environment, and that individuals from within the population that are best fitted to the environment reproduce more successfully and pass on to their descendants the characteristics that have made them successful. But – and this is a crucial key ingredient of the theory – the copying process by which individuals pass on their characteristics to their descendants is slightly imperfect, so that offspring are not *exact* copies of their parents. This allows for new characteristics to appear in a population, and to spread if they are successful (in the sense that they make individuals more likely to survive and reproduce successfully), but to be rigorously culled from the population, by natural selection, if they are detrimental – simply because they will be a hindrance in the competition to survive and reproduce.

None of this is in any doubt today. Biologists do sometimes argue about details of the way in which selection works, the nature of the copying process by which characteristics are passed on from one generation to the next, and the speed with which changes can take place. But the basics of our understanding of evolution by natural selection remain as Darwin spelled them out in the nineteenth century. Of course, Darwin himself had no idea how characteristics are copied and passed on from parent to offspring. That is where the DNA comes in.

A living organism such as a human being is built from the DNA recipe in forty-six libraries contained in one original cell. For human beings and other life forms which reproduce sexually, it happens that the special initial cell is produced by the fusion of two cells, one from each parent, which have each been produced by a special kind of cell division known as meiosis, and which each contain only twenty-three libraries of the recipe; this is why children inherit some characteristics from each side of the family. The forty-six libraries, in fact, consist of twenty-three pairs of libraries which are interchangeable, but not identical. Only one gene in each of a pair of matched chromosomes is actually used by the body (perhaps to decide eye colour). As we shall see, this is very important for evolution; but once a fertilized, single human cell begins to develop, the original recipe is copied faithfully every time the cell divides, so that every one of the 10^{15} (1 followed by 15 zeros) cells in your body, or mine, contains a perfect replica of the recipe used to manufacture the whole body.

The cell is still the basic unit of life on Earth, for human or bacterium. We do not know how the first replicators evolved, or how they came to 'invent' cells. But we know that cellular living organisms existed on Earth more than 3 billion years ago, and we can explain at least the broad outlines of how their descendants developed and evolved down to the present day. The selection of cells from one generation to the next depended on how well suited they were to exploit their environment at the time. This is what lies behind Darwin's concept of 'survival of the fittest', not 'fittest' in the sense of the most athletic, but having more to do with the way in which the pieces of a jigsaw puzzle fit together. Cells (or more complex creatures) which fit in particularly well with their environmental surrounding will reproduce more successfully than competitors of theirs who do not fit in so well.

The survivors today, including ourselves, are descended from a long line of successful reproducers. 'Fitness' in the Darwinian sense essentially means ability to survive and reproduce. 'Natural selection' (a term which Darwin used both to compare and contrast with the way stock breeders artificially select animals for breeding purposes) actually operates down at the level of DNA itself. The most successful individuals in evolutionary terms are those that do a good job of passing on copies of their chromosomal DNA to the next generation. The DNA itself is not 'interested' in the well-being of the body or cell it inhabits, except as a means of ensuring the spread of that particular DNA.

The way in which evolution has led to the diversity of life on Earth is worth going into in some detail, partly because we are all interested in our own origins, but also as an example of the power of this very simple process of Darwinian evolution operating by the selection of more successful variations on a theme from the less successful variations. All you need, remember, is an imperfect copying process to produce variety, and some form of competition to select from the variations on the theme. The familiar example is indeed the evolution of life here on Earth, but we shall see that there may be much more to the story than this.

After the appearance of the first replicator itself, the greatest invention of life has been the cell, a safe home which protects the replicator molecules from their surroundings. Life on Earth can be divided into two kinds, based upon two different types of cell, and the difference between the two types is the most profound division of all – far more important than, for example, the distinction between plants and animals. Both plants and animals are made up of the same kinds of cells, called eukaryotes. The name comes from Greek, and means 'true kernel', describing the most important feature of eukaryotic cells: the presence of an inner cell, or nucleus (the 'kernel'), which contains the chromosomal DNA. This is the kind of cell from which we are made.

Most cells are tiny (perhaps a tenth or a hundredth of a millimetre in diameter), although egg cells can be quite large, and the yolk of a hen's egg, for example, is a single cell. Although cells in a complex organism such as the human body may have different tasks to perform, all share certain basic features. The most important is the membrane that surrounds the cell, its barrier against the outside world. Although only a few ten-millionths of a millimetre thick, the membrane controls the environment inside the cell by allowing only certain molecules to get in (food) and certain molecules to get out (waste products). The membrane actively selects the molecules it allows to pass, 'recognizing' them by their size and shape so that chemical messages to and from the rest of the body can also pass in and out of the cell; inside, floating in a watery fluid called the cytoplasm, are a variety of specialized structures called organelles, which control the chemical processes by which food is converted into energy, messages are passed, and so on.

The nucleus is the central controller of all this activity, which, stretching the analogy, might be described as the brain of the cell. Its most important role is as a storehouse of information, the multiple 'library', which contains details not just of the workings of one cell, but of the whole body in which it resides, and of the cell's place in this grander scheme.

Among the organelles, mitochondria do the job of converting food molecules into energy, while ribosomes are responsible for the construction of new protein molecules from the available chemical raw materials. Plant cells, unlike animal cells, also contain structures called chloroplasts, containing chlorophyll, which are crucial in the process known as photosynthesis by which plants turn sunlight into energy. No animal can derive energy directly from sunlight, and all animals depend on plants (or on other animals which have themselves eaten plants) for food. In other ways, though, animal and plant cells are basically similar – and much more like each other than like the cells called prokaryotes (from the Greek, meaning 'pre-kernel').

Among the variety of life on Earth, prokaryotic cells make up only two families, the bacteria and another single-celled life-form sometimes called 'blue-green algae', but more usually known by the alternative name cyanobacteria, which I shall use. The cyanobacteria produce oxygen as a by-product of photosynthesis, like the plants, and this has played a vital part in the history of our planet. But both cyanobacteria and bacteria are single-celled species which reproduce simply by splitting into two; their cells have no organized nuclei, but merely a few strands of DNA (in the simplest cells, a single strand of DNA).

It is obvious why biologists class prokaryotes as 'pre' eukaryotes, and this simpler form of cell is much nearer to the form we would expect to have evolved from a chemical soup in which replicator molecules (DNA) had come into existence but were initially unprotected from the random chemical processes still going on in their environment. It is very likely that both mitochondria and chloroplasts are derivatives of what were once free-living organisms in their own right; Lynn Margulis, of Boston University, is perhaps the strongest proponent of the idea that modern eukaryotic cells developed from a combination of prokaryotic predecessors that learned to live together for their mutual advantage. Both mitochondria and chloroplasts contain fragments of DNA that resembles prokaryotic DNA, and the idea that eukaryotic cells developed from a combination of prokaryotic organisms is persuasively plausible, echoing the way many millions of cells can 'get together' (or, better, 'grow together'), in cooperating to make one animal or plant. It seems that earlier forms of life have learned to cooperate to make cells, while cells have cooperated to make larger organisms – and, indeed, in some striking examples (such as bees) individual creatures have 'learned' to cooperate on the next scale up, and all for the benefit of the survival of a few strands of DNA in their cells. Even this may not be the end of the story. Life can indeed be complicated, given the variety of niches open to life and the great time over which evolution by natural selection has been operating.

The earliest direct evidence of life on Earth comes from traces of the outlines of soft creatures, colonies of bacteria and cyanobacteria, in rocks more than 3 billion years old. But there is very little in the way of fossil evidence of life in any rocks more than about 600 million years old, and for all practical purposes the geological calendar begins with the sudden spread of a multitude of complex life forms, revealed by their fossil remains, after that time. This first and most important boundary in the geological calendar marks the beginning of the period known as the Cambrian. In geological terms, everything that went before, the first 4 billion years or so (about 90 per cent) of the history of the Earth as a planet, is called simply the Precambrian, and very little is known about it compared with our more detailed understanding of the past 500 or 600 million years.

But it was during the long Precambrian that life got its grip on the Earth, and evolution started life down the many-branched trail that was to lead to the diversity of species we know today. Geologists identify the boundary between Cambrian and Precambrian simply by the sudden (by geological standards) spread of life indicated by fossil remains in Cambrian rocks. In other ways the rocks are no different from the rocks of the immediately preceding Precambrian. It is life itself that defines this boundary between one geological division of time and the next. And the reason for the variety of fossils from the Cambrian onwards is that it was at that time that living creatures first developed easily fossilized and clearly identifiable features such as shells.

There was a great deal of life before the Cambrian, but it was soft-bodied forms of single-celled life, which left only microscopic fossil remains. Along the way those 'simple' life forms managed to have as dramatic an effect on the Earth as any subsequent form of life. They converted the atmosphere from the mixture of gases emitted by volcanoes, including plenty of carbon dioxide, into the oxygen-rich mixture of gases which we today and which then provided, as we shall see, a shield that was vital in allowing life to spread onto land, as well as the oxygen that is essential for the activity of animal life. All this took time, which is why life diversified and spread so explosively only after the stage had been suitably set in the Precambrian. Fossil remains of jellyfish, worms, sponges and the like have now been identified in rocks dating back to about 100 million years before the end of the Precambrian; it is in times still earlier than this (more than 700 million years ago) that the single-celled life forms can truly be said to have dominated the Earth.

The microfossil remains clearly indicate that prokaryotes came first, and that the larger cells typical of eukaryotes arrived later on the scene. But the different abilities of the two kinds of cell already reveal that this must have been the case, since eukaryotes almost all require oxygen in order to live (even the few odd eukaryotes which do not require oxygen seem to have evolved from ancestors which did need oxygen), whereas prokaryotes of one form or another show a wide variety of oxygen requirements. Some bacteria cannot grow or reproduce if any oxygen is present; others tolerate oxygen but can get along quite well without it; and there are prokaryotes which do not actually need oxygen but reproduce best if there is a little around (less than the concentration in the atmosphere today), as well as some who cannot manage without oxygen. This is just the pattern we might expect if prokaryotes had diversified into different single-celled species at a time when oxygen was slowly building up in the atmosphere. And the absence of such diversity in the oxygen needs of eukaryotes tells us that they all developed only after the oxygen concentration of the atmosphere had reached something near its present level. All of the evidence confirms that prokaryotes were indeed 'pre' eukaryotes and remain the earliest identifiable native life-forms of planet Earth.

One all-important cell process in eukaryotes cannot proceed at all without oxygen. No eukaryotic cell can divide and copy itself (the process known as mitosis) unless there is at least a little oxygen present, which means that the eukaryotic single-celled organisms could not reproduce without oxygen – surely the clinching evidence that they evolved after the atmosphere had been transformed. How, then, was the atmosphere transformed – and when?

We can get a good idea of the time oxygen appeared in the atmosphere of the Earth by looking at the Precambrian fossil evidence of eukaryotic remains. Some fossil filaments similar to modern fungi and green algae have been found in Siberian rocks about 725 million years old; eukaryotic microfossils from the eastern Grand Canyon have been dated as some 800 million years old, with a similar age for some Australian algae microfossils. So oxygen-breathing eukaryotic life was firmly established by 800 million years ago.

Further back in time, remains that look very like eukaryotic cells have been found in rocks as much as 1.5 billion years old, and these include microfossils so well preserved that it is possible to see the outline of what seem to be organelles within the cells. But although many microfossils older than this have been found, none of them can be definitely identified as eukaryotic. About 1.5 billion years ago there is a clear break in the microfossil record, as significant as any distinction between geological periods determined from changes in the macrofossil record.

From about 3 billion to about 1.5 billion years ago, prokaryotes ruled the Earth. Then came eukaryotes, to be followed by all the species that could thrive on a planet with an oxygen-rich atmosphere. And the biological processes that produced the oxygen, and continue to recycle it today, must have started their work sometime between 3 and 1.5 billion years ago. Once again, so-called primitive forms of life (which have, in fact, survived efficiently for 3 billion years, so they must be doing something right) still exist on Earth today, doing the same job that their (and our) ancestors must have done back when there was no free oxygen around, getting energy out of sunlight by photosynthesis.

In the first step of photosynthesis, light energy is absorbed by molecules sensitive at particular wavelengths, in the exact reverse of the process by which those molecules would emit energy at distinct wavelengths, producing bright lines in the electromagnetic spectrum, if they had extra energy to get rid of. The solar energy is then used to drive a series of chemical reactions beneficial to the organism which is doing the photosynthesis. Without photosynthesis, it would not be just green plants we would be short of – there would be no animals either, since animals cannot convert sunlight into energy in the form of metabolizable compounds, and we are completely dependent on plants for food, whether we eat plants directly or eat other animals that have eaten plants in order to live and grow.

But why *green* plants? The colour depends on exactly which wavelengths of sunlight are being absorbed, and green represents the surplus radiation which is not absorbed but is reflected back as visible light. Green plants use molecules of chlorophyll to do their energy absorbing, using light in the red and blue–violet parts of the spectrum and leaving most of the yellow and green sunlight to be reflected away. This is very curious, because our Sun radiates a great deal more energy in the yellow–green part of the spectrum than in the red and blue–violet, and there are other compounds which could be used in photosynthesis much more efficiently than chlorophyll. Indeed, some plants have adapted to use other pigments, reflecting away the red light and absorbing high-energy yellow-green light so that they look red. We know that these plants have evolved from photosynthesizers that used chlorophyll because they still use chlorophyll as well: the energy absorbed by the 'new' pigments is passed *first* to chlorophyll – a completely unnecessary step – and then on down the chain to where it is needed.

Two important things (one definite, one speculative) can be learned from this. First, such a pattern, in which later evolutionary adaptations are tacked onto already existing systems, is typical of the way evolution works. At a grosser level, in animals the organs that were fins in fishy ancestors have been adapted to become legs and arms; the land animals did not lose their flippers and then 'invent' arms and legs.

This pattern of change is particularly clear from the study of foetuses, which shows that every human being develops in the womb through stages which are very like fish, reptiles and non-primate mammals before becoming distinctly human in appearance. This recapitulation is a consequence of the fact that we are descended from ancestors of all these forms, each of which developed from an egg. Each change is made by adding something new to the process of development from the egg. At each stage there is enough leeway for change in the DNA code to produce a slightly different 'model' of what went before, like the difference between a 1992 car and its slightly modified 1993 version. But the complex web of interactions that keeps a living organism alive would break down if wholesale changes were made, like ripping out the engine and sticking a steam engine in its place.

The second and more speculative interpretation of the curious way in which plants on Earth seem to have adapted for light different from the light of our Sun is that photosynthesis may not have been invented on Earth. If Hoyle and Wickramasinghe are correct, then even photosynthesizing cells may have developed in space before life reached the Earth, and perhaps chlorophyll is a much more appropriate pigment for the efficient absorption of light under those conditions. Hoyle and Wickramasinghe themselves have pointed out that the chemical rings which form a basic part of the chlorophyll molecule (each ring containing one nitrogen atom and four carbon atoms, C_4N) are members of the porphyrin family. The properties of this chemical family can exactly explain one particular feature of the absorption of light in interstellar space observed by astronomers, a 'line' at 443 nm wavelength. The line would be produced directly by the absorption of light by these chemical rings. There may be other ways to explain this particular interstellar absorption line, and theirs is still a controversial interpretation of the evidence; but they say (in their book *Lifecloud*) that 'the basic constituents of chlorophyll may therefore well have been added to the Earth' from space after our planet formed.

If they are right, that would certainly be a good reason why life on Earth should have started photosynthesizing with the aid of chlorophyll rather than some other compound. Once it had started photosynthesizing, of course, competitors could never have set up in business since all organic material was rapidly becoming locked up in living cells, and any improvement to the system – such as the addition of red pigment – had to come later. Back at the beginning of the story of life on Earth, though, the first photosynthesizers differed from most of their present day descendants in one crucial respect. They did not 'throw away' the oxygen produced as a waste product of photosynthesis, but carefully packaged it together with other molecules as a safe, unreactive compound before ejecting the waste from the cell. To those first living cells oxygen was a poison, reacting so violently and quickly with organic compounds that it would disrupt the metabolism of any cell that let it go in its pure form.

When the first cells that learned to live with free oxygen appeared, they had a huge advantage over other life-forms. They no longer had to go through the energy-consuming business of packaging their waste oxygen before throwing it away and, as a bonus, the oxygen they threw away was actually harmful to the other cells around which had not learned to tolerate it. Once the oxygen producers appeared, they must have established themselves very quickly across the oceans of the world. Several lines of evidence show that the major transition from an oxygen-free atmosphere to one containing at least 1 per cent as much oxygen as it does now took place within a few hundred million years, about 2 billion years ago.

The best evidence comes from widespread deposits of iron oxides, found everywhere in the world today. These are known as the banded iron formations, or BIFs, and are all between

1.8 and 2.2 billion years old. Iron can form several different compounds with oxygen, but not all of them are soluble in water. When a little oxygen is present, iron forms a soluble 'ferrous' oxide; when more oxygen is available it forms the insoluble 'ferric' oxide. The explanation of how the BIFs came to be laid down is that in the original oceans of our planet, with no free oxygen around, iron was dissolved in its ferrous state. When oxygen became available, it triggered a very well-known reaction called rusting, in which all the ferrous iron in solution was converted to insoluble ferric oxides and settled in thick layers over the ocean bed. When oxygen appeared on Earth, all of the iron rusted. To do the job in such a short time – a few hundred million years – there must have been a steady supply of oxygen from some new source, and only aerobic photosynthesis (photosynthesis that releases oxygen) can explain the pattern of events.

The concentration of oxygen in the atmosphere could begin to build up only after the BIFs were deposited, since before all the iron had rusted oxygen was being trapped in rust as quickly as it was being produced. This completes the story of the Precambrian, and of the influence of Precambrian life on our planet, as far as we know. The first photosynthesizing organisms, anaerobic prokaryotes, were around at least 3 billion years ago. The invention of aerobic photosynthesis a little more than 2 billion years ago gave the aerobic organisms an advantage and must have wiped out almost all the earlier forms of life, while first rusting the entire ocean and then transforming the atmosphere (an early example of global pollution). Eukaryotic cells developed about 1.5 billion years ago in a stable environment rich in oxygen, after this transformation had been completed, and rapidly diversified. By about a billion years ago, sexual reproduction – a key invention, as we shall soon see – had appeared, and over the next 400 million years even more diversification took place, producing many distinct multicellular species by the end of the Precambrian.

Since then, change has been rapid. This is partly because of the energy available to organisms which can use oxygen in respiration, energy which allows them to invade new ecological niches and to compete with one another (in the evolutionary as well as the everyday sense) for food. It is also because of the invention of sexual reproduction, which allows for the spread of diversity among the 'gene pool' of a species. And it is because, under the protecting umbrella of ozone in the new oxygen-rich atmosphere, life was able to take to the land, a whole new habitat with a whole new array of selection pressures operating on species.

Before concentrating on the biological reasons for this diversity of life on Earth today, it is appropriate to look at the nature of the atmosphere we have inherited from the Precambrian, an atmosphere not just rich in oxygen but with its own layered structure, protecting us from the Sun's harsh ultraviolet radiation, which would otherwise render the land masses of our planet uninhabitable. Life changed the Precambrian environment as much as any physical process did, and we are living off the benefits provided by our cyanobacterial ancestors 2 billion years ago.

The structure of the atmosphere may be described most simply in terms of temperature. Heated by incoming solar energy which passes through the atmosphere largely without being absorbed, the ground is warm and radiates heat at infrared wavelengths back out towards space. But this infrared heat *is* absorbed by water vapour, carbon dioxide and other gases in the air. So some of the outgoing heat is trapped, by what is known as the greenhouse effect; also, some incoming solar radiation is reflected by clouds, snow, and the land or sea surface. Overall a balance has been struck – although there are minor variations in the balance from time to time, sufficient to produce the pattern of repeating ice ages and warm epochs characteristic of the past few hundred million years.

The lowest layer of the atmosphere is called the troposphere. The temperature in the troposphere falls off with increasing height above ground, initially by about 6°C for every kilometre of altitude gained. The decrease slows near 10 km (6 miles) altitude and stops near 15 km (9 miles). From about 20 to 50 km

(12 to 30 miles) the temperature increases with altitude, from a minimum of about −60°C to about 0°C maximum at the top of this warming layer, which is called the stratosphere. Warming indicates that energy is being absorbed in the stratosphere, and the molecules which do the energy absorbing are those of ozone, a form of molecular oxygen with three atoms per molecule (O_3) instead of the usual two (O_2).

In the stratosphere, ozone is produced by a series of dynamically interacting chemical reactions driven by sunlight (photochemical reactions). Ordinary diatomic molecules of oxygen are split into their component atoms as they absorb ultraviolet energy from the Sun, and the efficiency with which free atoms of oxygen are produced depends on a balance between the number of molecules there are that can be split (more at low altitude) and the amount of ultraviolet energy available (more at high altitude). Once free oxygen atoms are produced, they combine with other diatomic molecules to make ozone. Because of the factors affecting the photochemical reaction rates, the ozone concentration is greatest in a band of the stratosphere between about 20 and 30 km (12 and 20 miles) altitude. The ozone, especially in that band, also absorbs electromagnetic waves of wavelength below 280 nm, stopping this ultraviolet radiation from reaching the ground.

Ozone is constantly being broken down into diatomic oxygen and single oxygen atoms, but the ozone layer is also constantly being replenished. Although the individual molecules do not stay in one state, a rough overall equilibrium is maintained, rather like a bucket with a hole in it being filled from a tap: water is always running into the leaky bucket and always running out, but the level of water inside may stay much the same.

The array of interacting processes which maintain the ozone layer is affected by changes in solar radiation, so that the concentration of ozone varies from day to night, with the cycle of the seasons, and over the Sun's roughly 11-year cycle of activity (the sunspot cycle). It is also affected by the presence of other chemical elements and compounds in the stratosphere, and both chlorine and some nitrogen oxides are very efficient at shifting the equilibrium so that the concentration of ozone in the stratosphere is reduced. Harmful effects on the ozone layer of the stratosphere are produced by pollution from chlorine released by the breakdown in the stratosphere of the chlorofluorocarbon (CFC) gases used in some spray cans, in foamed plastics, as the working fluid in refrigerators and in other applications. They have caused the now notorious 'hole' which appears in the ozone layer over Antarctica each spring, and the lesser, but growing, damage to the ozone layer over the Arctic.

Above the stratosphere there is another cooling layer, the mesosphere, and at the top of this, at about 80 km (50 miles) altitude, the temperature is a chilly −100°C. From here on up – or out – temperature is no longer a good guide to conditions in what is left of the atmosphere, and by 500 km (300 miles) collisions between atoms and molecules are too rare for it even to be thought of as a continuous gas. Instead, these outer fringes are described in terms of their electrical properties, on the extent to which atoms are ionized. For us, though, what matters is the troposphere in which we live, and the stratosphere immediately above which acts as a lid to the troposphere (because the warming layer stops convection, so that clouds and weather occur only in the troposphere below) and shields us from ultraviolet. The crucial importance of this shield is clear from the fact that ultraviolet radiation is widely used as a sterilizing agent for surgical instruments, which have to be free of bacteria and other micro-organisms. The electromagnetic energy in this waveband is particularly disruptive to DNA – the 'resonance' set up as DNA molecules absorb certain wavelengths of ultraviolet radiation is strong enough to damage the molecules. This is why solar ultraviolet radiation is implicated in skin cancer, which is a result of mistakes in cell division and growth produced by faulty DNA replication. The DNA of the earliest life-forms, lacking a thick skin, would have been particularly susceptible to this kind of disruption, so that life had to develop first in the sea (which, in any case, is where all the organic molecules were dissolved). Perhaps a thick-shelled creature of the shallow waters, impervious to ultraviolet, might eventually have colonized the land, although it would have found nothing to eat there; but once the ozone layer was established the hazard was removed and relatively thin-skinned plants were able to spread onto the land, to be followed by animal life.

The speed with which life diversified after about 1.5 billion years ago clearly owed a great deal to the presence of oxygen in the atmosphere and the invention of respiration as a means of obtaining energy. But the spread of life into different ecological niches to produce the diversity we see around us today also depended to a very great extent on another development, sexual reproduction. If you look at the variety of life around you, almost everything you see reproduces sexually, with two different types of parent required to produce a new generation of copies of the organism. Asexual reproduction, in which one creature produces, on its own, an exact copy of itself, largely remains the province of single-celled creatures like our Precambrian ancestors. Clearly, sex is a great advantage in the natural selection game, and has been for the past billion years or so. Fossil remains of the earliest known animals – jellyfish, worms and corals – are dated to around 650 to 700 million years ago, but the complexity of these multicellular organisms clearly shows that they were products of evolutionary processes going back hundreds of millions of years before that.

We do not know exactly how or when the first sexually reproducing multicellular ancestors of what are now land animals appeared, although we can certainly make a good guess at both how sexual reproduction developed and why it was successful. But we do know that, as well as the biological factors affecting the rate of evolutionary change and the diversification of life at the end of the Precambrian and into the Cambrian, there were changes in the physical environment which must have played a part by the evolutionary pressures they produced.

To understand why sexual reproduction has proved to be such an advantage, we have to look again at reproduction at the molecular level, where DNA strands are copied and passed on from one generation to another. One molecule of DNA, with its spiralling double strands matched by chemical bonds linking particular pairs of molecules, splits down the middle to produce two single strands. Each of the two strands can then quickly rebuild a whole double helix by selecting nucleotides from the biological material around it in the cell to pair up with the 'broken' bonds. The four compounds which make up the four-letter DNA alphabet (A, C, G, T) form bonds in only two ways: A always with T, and C always with G. So, if the double helix is unzipped, one broken bond may leave A at a particular site on one molecule and T at the corresponding site on the other. The A will recombine only with another T, and the T only with another A, so that the two single strands, each opposite halves of the whole, rebuild themselves into two new DNA molecules, each identical to the original. It is rather as if a pair of gloves, separated and thrown into a heap of odd gloves, each paired up with another glove to make a complete set, the left finding a new right one and the right finding a new left one. But it is rather more complex with DNA, where millions of nucleotides have to be untwisted, unzipped, paired up and put back together again – in the case of a bacterium, all within about twenty minutes, the time it takes for a cell to divide.

When DNA is passing messages about the workings of the cell or body it inhabits, the process is a little different. Then, only part of the molecule unzips, leaving a free loop of DNA with the bases of its four-letter code exposed along a particular sequence, spelling out a particular message. This message is used to control the construction of specific protein molecules used by the machinery of the cell.

If the DNA passed on to a next-generation cell contains a copying error, then the offspring will be different from its parent. Such mutations are usually harmful: the offspring dies or reproduces less successfully as a result, and that is the end of the story. Occasionally, though, the mistake is an improvement, enabling the offspring to reproduce more successfully than its parents and other relations – in which case the parental strain is the one that dies out or becomes pushed aside into a specialized ecological niche where its old-fashioned ways will still work to advantage.

We have in our cells today a great deal more DNA than our single-celled ancestors which were alive in the Precambrian. But the rules of reproduction are much the same regardless of how much, or how little, DNA the cell has, so once again it makes sense to look at modern examples to find out how evolution works, in the confident expectation that the game has been played by the same rules for as long as there has been life on Earth. One gene may carry a very simple message, such as the gene for blue eyes, but in general each gene will affect parts of the body in different ways, and each part of the body is constructed in accordance with a combination of instructions from many different genes. This, combined with the great many possible combinations of genes that sexual reproduction allows, is why individuals are so different from one another. For simplicity, though, we can think of genes as the basic components of the code describing how to build a body (the genetic code), and we can think of each gene, or group of genes, as carrying a simple message such as 'blue eyes', 'long legs' or 'brown skin'. Mistakes in copying genes – genetic mutations – happen all the time, since although there is enormous selection pressure for copying accuracy, there is also an enormous number of copies of each gene around today. With 4 billion human beings on Earth, for example, virtually every gene appears in a mutated form in one human body or other, even though almost 4 billion perfect copies of every gene are also present in the overall human gene pool. Or, rather, not quite 4 billion copies of each *gene*, since there may be several different versions of the gene for, say, eye colour.

Each human being carries forty-six chromosomes, twenty-three inherited from each parent. So each human being has two sets of genes, and the gene for eye colour inherited from the mother might say 'brown', while the gene for eye colour inherited from the father might say 'blue'. In that particular case, the human being built from that set of DNA will not have one brown eye and one blue; the blue gene is 'recessive' and the brown 'dominant', so that in practice both eyes will be brown and the gene for blue eyes is ignored. Such competing genes, which offer different ways of doing the same particular bit of body building, are called alleles, and in the eye colour example there are other possible alleles in the human gene pool, although any one individual can only have two such alleles, one inherited from each parent. Of course, both alleles may say the same thing – both instructing for blue eyes, perhaps – in which case there is no conflict. Each allele has been produced by mutation from a previous version of the gene, and it is quite possible for a large number of mutated genes, alleles, to exist in a population of a particular species.

This is a key feature of the workings of evolution. Continual small changes in the genes provide for variety; then, if circumstances change so that some particular allele is favoured, it will spread rapidly through the gene pool, displacing its rivals because the bodies they live in die young or fail to reproduce successfully. A new population, a variation on the previous form of the species, becomes established, and then the slow processes of evolution build up variations on the new theme in the form of new alleles. Mutations do not happen suddenly, producing dramatic physical changes in the body of a new individual, compared with the bodies of its parents. Nor do mutations happen in response to environmental changes – the soft-bodied animals of the late Precambrian did not 'know' that it was getting colder, or that there were more predators about, and grow their shells in self-defence. Rather, they must have carried an allele for thicker skin, competing with an allele for thinner skin. When the climate changed or predators spread across the seas, the thin-skinned individuals were killed, and only the possessors of the thick-skin allele survived. Repetition of this process over many generations produced creatures with hard shells.

Taking a hypothetical human example, even though the allele for blue eyes is recessive, it is widespread in the human population and may be present even in people with brown eyes. Suppose that some change in the nature of the radiation from the Sun had made blue eyes an advantage when our ancestors depended on hunting and gathering to make a living. Then 'blue-eyes' would have spread very quickly – the 'brown-eyes' might die of starvation because they could no longer see well enough to hunt. Taking things a step further, a brown-eye who had a blue-eyed son might survive if the son caught enough food for them both, while a brown-eye who had a brown-eyed son would starve along with him. The more genetic variation there is within a population, the quicker it can adapt to changing circumstances. And sexual reproduction is the most efficient way of ensuring that there is a great variety of genetic variation within a population.

When cells divide to produce a growing body, all forty-six chromosomes (or, more accurately, each of the twenty-three chromosome pairs) are copied precisely into a duplicate set of chromosomes, one set going into each of the cells produced by the division. This process of cell division, mitosis, is almost identical to the way a single-celled organism makes copies of its DNA before the cell splits in two – the main difference is that while those two cells then go their separate ways, in a multicelled creature the two cells stick together and help to make up a greater whole. But in the first stage of sexual reproduction a very different form of cell division, meiosis, takes place. In meiotic cell divisions, whole chunks of chromosome are detached and swapped between pairs to make new chromosomes which contain the same genes as in the parent, but arranged in different combinations of alleles. The cell then divides, without copying the chromosomes, in a two-stage process which produces sex cells (sperm or eggs in animals) which each contain only one set of twenty-three chromosomes. So, for example, you might inherit a blue-eye allele from your father and a long-leg allele from your mother, on separate chromosomes, but can pass both of them (along with many other alleles) on to your child on a single 'new' chromosome formed during meiosis.

Because chromosomes are so long and contain so much genetic material, this process of cutting chunks out of one chromosome in a pair and swapping it with material from the corresponding paired chromosome produces an enormous variety of new chromosomes, and there is virtually no chance that any two sperm or eggs produced by one individual will carry identical chromosomal recipes instructing how the new individual should be built. The swapping process is aptly termed 'crossing over', and the common analogy is with shuffling a very large pack of cards, although a better analogy would be shuffling two packs of cards and then swapping cards between the two packs. And it ensures that every generation throws up new patterns of genetic arrangements from the variety present in chromosomes. Among other things, it means that although, in principle, you could examine the material from one of your own cells and identify the twenty-three chromosomes which come from your mother and the twenty-three from your father, similar examination of the twenty-three chromosomes in one of your sex cells would show that none of them could be identified as coming from one of your parents as opposed to the other. They all contain chunks of DNA from both your parents, the grandparents of the new human being that could develop if the cell were to fuse with one of the opposite sex.

This, of course, is the next stage of sexual reproduction. The organism cannot reproduce on its own, but must find a member of the opposite sex with which to mate. When that happens, two sex cells, each containing twenty-three chromosomes, fuse to produce one cell with forty-six paired chromosomes. The process of building a new body can then begin, with the specific details of each stage of the construction being read off from one member of a pair of chromosomes according to whether the allele for that particular stage of development is dominant or recessive on one chromosome or the other.

There are obvious disadvantages to this method of reproduction – just finding a partner may be difficult; indeed, the need to search for one may make the organism susceptible to dangers that it could avoid by hiding and reproducing asexually. It also means that each parent contributes only half the blueprint for the next individual, instead of ensuring that all its genes are passed on intact. And all this breaking up of chromosomes and crossing-over of DNA must introduce the chance of mistakes arising in the copying process.

But the last of these may actually be an advantage, provided there is enough variation to ensure a flexible response to changing environmental conditions, or to the new problems being posed by the evolution of new rival species, but not so much variation that the offspring can no longer reproduce. And, clearly, the advantages of sexual reproduction must dramatically outweigh the disadvantages, or I would not have written this book and you would not be reading it. Variability is surely the secret of the success of sexual reproduction, the variability ensured by all that crossing over, shuffling of genes and provision of two sets of chromosomes, with many alternative alleles, from two parents.

Sexual reproduction is the best way to spread genes in a diversity of species into a new and changing environment. That is exactly the situation that existed in the late Precambrian and early Paleozoic – the atmosphere of the Earth was changing, with oxygen concentration increasing; the climate changed, into and out of an ice age; and the continents themselves were moving, breaking up and forming new patterns.

But although whole species may benefit from the process of sexual reproduction, it is important to remember that the nitty-gritty of selection is at the level of the individual. It is an individual that lives or dies, reproduces or fails to reproduce, and it is the individual's genes that are (or are not) passed on to another individual (or individuals) in the next generation.

The clinching evidence that this is the basic mechanism of selection in nature comes when numbers are put in. It is fairly straightforward to produce equations describing the behaviour, in mathematical (statistical) terms, of large numbers of individuals obeying well-defined rules, whether those individuals are animals giving warning calls, or involved in mating displays, or molecules which make up a gas. John Maynard Smith, of the University of Sussex, has developed models to determine just which pattern of behaviour 'ought' to dominate an animal community in different sets of circumstances. The branch of mathematics used by him and his colleagues is called games theory, and it has a very sound basis thanks to the investment of effort over the years in various attempts to simulate war and predict the outcome of a chosen strategy of attack or defence before putting it to the test. At the heart of Maynard Smith's application of games theory to animal behaviour is the idea of an 'evolutionarily stable strategy' (ESS), a pattern of behaviour which will persist in a population, though individuals may come and go, for many generations – because any other strategy which some individuals in the population might develop cannot do as well as the ESS.

This is particularly important because the games theory approach treats individuals as individuals, doing what is best for them. There is nothing here about the good of the species; individuals act for the good of themselves. What emerges, time after time, is clear evidence that species in the real world, following patterns of behaviour that look at first to be tailored to the good of the species as a whole, are actually doing what is best for each individual, and are meshed in to the workings of an ESS. Richard Dawkins, in his book *The Selfish Gene*, does full justice to these ideas and explains in detail how the ESS works in a variety of cases; by showing how a collection of individuals can appear to be working for the common good in line with some grander scheme of things, this provides the most important advance in evolutionary thinking since the time of Darwin. So here is one of Maynard Smith's classic examples of the ESS at work, the Hawks versus Doves scenario.

Imagine a population of animals of one species, with each individual either a Hawk or a Dove in the sense used to denote aggressive or peaceful behaviour in humans. The Hawks always fight when they meet a rival; the Doves may threaten, but always run away if an opponent attacks. When a Hawk meets a Dove, no one is hurt because the Dove runs away; when a Dove meets a Dove, no one is hurt because after some mutual threatening they both run away; but when a Hawk meets a Hawk they fight until one is severely hurt. If we assume that conflicts between individuals arise over something of material value – food, or the opportunity to mate, perhaps – it is possible to allot arbitrary 'points' for success or failure in the competition of individuals. This is a purely hypothetical example, so for convenience the numbers can be 50 points for a win, 0 points for running away, −100 for being severely injured and −10 for wasting time in a mutual threat display. The points represent a direct measure of genetic success: individuals who score most get the most food and the most opportunities to reproduce, so their genes survive into the next generation. Low-scoring individuals have less chance of propagating their genes. The question which games theory can answer is whether there is an ESS in the Hawks versus Doves scenario and, if so, what is the stable balance of Hawks and Doves.

The first thing we learn is that a population of all Doves or all Hawks is not stable. Take Doves first. If all the population are Doves, then every time two come into conflict they have a threat display and each scores −10 points. But one runs away first, so the other scores 50 points by picking up the reward that was in dispute, getting a net score of 40. If each individual wins half the contests and loses half, the average score per individual per contest is 15 points, the average of 40 for a win and −10 for a loss. Things look pretty good in an all-Dove society, and nobody gets hurt, starves or fails to reproduce. But now imagine that a genetic mutation produces one single Hawk among the Doves. The Hawk wastes no time threatening, but just chases Doves away, scoring 50 points every time there is a conflict. So his average score is also 50, and he is doing vastly better than all the Doves, still averaging only 15 points. Hawk genes must spread rapidly through the population as a result, until there are a significant number of Hawks around and they start to come into conflict.

At the other extreme, imagine an all-Hawk society. Every time two of them meet they fight bitterly. One is severely injured and scores −100; the other wins and scores 50. But the average score is a measly −25, the average of 50 and −100, and any Dove mutation that occurred in the population, having a comparatively better score of 0 thanks to his cowardice in running away from trouble, will do better than the Hawks, at least until there are enough Doves around to provide easy pickings for the remaining Hawks.

Clearly, the ESS lies somewhere between the two extremes. For these particular numbers, the stable population consists of five-twelfths Doves and seven-twelfths Hawks – seven Hawks for every five Doves. In the real world this is the same as saying that the stable *strategy* is for each individual to behave like a Hawk seven-twelfths of the time, and like a Dove five-twelfths of the time, without giving any warning in advance of which course of action it would follow in any particular conflict. Genes which carry the orders, in effect, 'be aggressive a bit more than half the time, but run away almost half the time' are successful and will spread to produce a stable population following a stable strategy. The fact that this has nothing to do with the good of the species is borne out by the amount of success this particular ESS produces – on average, in the seven Hawks to five Doves mixed society, each individual scores 6.25 in each conflict. This is much less than the material advantage that can be gained by each individual and by the species as a whole in an all-Dove society (15 points per conflict). The species would do a lot better if all the individuals were Doves, but this cannot happen because sooner or later one Hawk will arise by mutation and be enormously successful as long as there is a majority of Doves around. The aggression gene in the Hawk cares nothing about the species; it cares about producing replicas of Hawk genes and the success of individual bodies at making replicas of Hawk genes.

Even such a simple example raises intriguing speculations about the ESS for human individuals. Do we carry 'Hawk' genes that are detrimental to the species as a whole in the same way? Is it possible that the relatively new evolutionary development of intelligence gives us a chance to work out the numbers and begin to act for the good of the species rather than the good of the individual, producing benefits for the individual as well, in the equivalent of an 'all-Dove' society? Such questions are beyond the scope of this book (I have looked at them in more detail, with Mary Gribbin, in *Being Human*). But the example highlights the way in which blind evolution, selection operating at the individual level on the variety of individuals produced by variation within a population (variation which is particularly likely to occur when the population reproduces sexually), can lead to complex patterns of behaviour which turn out to follow simple mathematical rules. There has to be variation – something for selection to act on. And there has to be an advantage of some kind for one variation on the theme over another, or there would be no basis for selection. But, given variation and selection, and a long enough time span, evolution can produce human beings out of single-celled bacteria.

Why should the process stop there? When we trace the ancestry of humankind forward from the earliest cells on Earth to the present-day, the story inevitably unfolds as if evolution were working to produce a specific end-product that is better than what went before. To many people, humankind is still seen as the end-point of evolution, a 'superior' creature compared with all the other products of evolution. But evolution has not finished, and there is no reason to think that we represent its end-point; nor are we superior, by any biological or evolutionary standard, to other species – just different. All species that are alive on Earth today can be regarded as successful, and even species that are now extinct were successful in their day.

If we try to overcome our natural human chauvinism, we can see that the important thing about life on Earth is *all* of life on Earth – the fact that there is any life here at all. There is a web of life, in which we depend on plants that convert carbon dioxide into oxygen and provide our food, plants depend on bacteria that 'fix' nitrogen from the air and deposit it in the soil in a natural process of fertilization, and so on. As far as we know, there is no life at all on our nearest neighbour planets, Venus and Mars, while the Earth swarms with life; it seems that you cannot have a little bit of life on a planet, but that life is an all-or-nothing affair, which either 'infects' a planet completely or does not take a grip at all. This new perspective, on the Earth as a living planet rather than merely an abode of life, is one basis for the Gaia hypothesis, which has transformed ideas about the nature of life itself and the relationship between life and the cosmos. The new ideas do not overturn our established understanding of how evolution works, but they do provide a stepping stone to take those ideas out into the broader context of the Universe at large.

CHAPTER FIVE

The Living Planet

The extraordinary extent to which living things cooperate to their mutual benefit is clear when we look in detail at the structure of a typical eukaryotic cell. Such cells may be small, but do not make the mistake of thinking that small means simple. There is so much activity going on inside them, with different parts of the cell specialized for different jobs, that one of the closest analogies you could make would be with a city.

In a city, there will be power-stations that provide the energy needed to keep the community running (and in a modern city some of the power might come directly from sunlight); there will be a 'nerve centre', at City Hall, which provides the instructions to operate the various systems of the city, including transportation and refuse disposal; there will be ways to bring raw materials, including food and fuel, into the city, and ways to get waste-products out again.

In a cell, there are analogues of all these activities. Distinct parts of the cell are given over to power generation, either using chemical energy from raw materials imported into the cell or using sunlight directly in photosynthesis first. The cell's nucleus, containing the cell's DNA, controls the workings of the cell with chemical messengers. And cells can do things that cities cannot. Virtually all cells are capable of reproducing, copying themselves complete; very many cells can also move around, swimming with the aid of hair-like appendages that thrash away at the fluid in which the cell floats. Yet all of this activity takes place in a tiny – literally microscopic – volume of living material.

Cells are little bags of biological material (less than a tenth of a millimetre across), a jelly-like substance contained within the membrane of the cell wall, like a blob of blancmange in a little plastic bag. The DNA inside the cell, so important for the cell's life and reproduction ('City Hall'), is just one component of the cell's life system. There are various chemical bits and pieces that enable the cell to turn the DNA message into action, and there are also structures so distinct from the rest of the cell 'jelly', like cells within cells, that the best explanation of their presence is that that is indeed what they are. These semi-autonomous 'cells within the cell' are called organelles. As we saw in Chapter Four, Lynn Margulis has explained how the ancestors of the organelles used to be separate, bacteria-like life-forms in their own right, arguing that eukaryotic cells have evolved by the successive incorporation of these outsiders into ever more complex cell structures, in a kind of bacterial symbiosis.

There are three kinds of organelle found in eukaryotic cells. One is a kind of thrashing tail (or tails), on the outside of a cell, which enables it to move around. This is very important for a cell that lives on its own. Although most of the cells of a human body have no need to swim about independently, and so have no external undulipodia, as they are termed, one special kind of human cell, the sperm, has a very efficient tail. The sperm's tail is essential in its quest to find and fuse with a human egg cell. Without those undulipodia, none of us would be here.

The second kind of organelle, the mitochondria, are the chemical power-stations that release the energy the cell needs. That stored-up chemical energy itself originally comes from sunlight, and is captured in the cells of photosynthesizing plants by the third kind of organelle, the chloroplasts or plastids, which is where photosynthesis takes place.

People and other animals do not have plastids in their cells, but obtain their energy supplies by eating plants, or by eating animals that have themselves fed on plants. What is more, all of the fossil fuel (coal, oil and gas) that we burn in our homes, power-stations and vehicles came originally from living things that died long ago, and whose remains were buried in the rocks and converted into coal, oil and gas by heat and pressure over millions of years. It is all carbon originally fixed out of carbon dioxide in the air by photosynthesis; when we burn fossil fuel we are releasing stored sunlight. So although we do not have plastids in our own cells, we could not live without them, any more than we could live without the mitochondria that release the stored energy when it is needed.

The most complex eukaryotic cells seem, therefore, to have evolved from the symbiotic getting-together of four separate ancestral lines – the 'host' cell, undulipodia, mitochondria and plastids. Some cells include all four components. In other cases (the cells of the leaf of a tree, for example) there is no need for external undulipodia, but all the other components are present; in most of the cells of the human body, there is no need either for external undulipodia or for plastids; but the basic story is still the same.

But how could quite separate kinds of primeval cell have 'learned' to cooperate in this way? Surely the whole idea flies in the face of the Darwinian concept of 'survival of the fittest' and competition for scarce resources? Not at all, because life is not a matter of beating your opponent, but of surviving yourself; if two organisms each survive more effectively – and leave more offspring – by working with each other, then that in itself makes the combined organism 'fitter', in the Darwinian sense, than either of its components working alone. In evolution, there does not have to be a loser for every winner. There can be more than one winner, and there can be more than one loser, and the two numbers can be different from each other. In the jargon of games theory, life is a 'non-zero-sum game', unlike, say, chess, which *is* a zero-sum game because for every winner there must be precisely one loser.

This would all still seem a little far-fetched, though, were it not for an extraordinary discovery made by Kwang Jeon, of the University of Tennessee, who spotted this kind of evolution at work in his laboratory, and gave it a helping hand. Jeon was working with microscopic, single-celled creatures called amoebas, when they were struck by a disease that spread through all the amoeba colonies in his laboratory. Almost all of them stopped feeding, ceased dividing into two to reproduce, and eventually died. Just a few of the amoebas seemed to be struggling along in spite of the disease, dividing about once every month instead of once every day, as they had done when they were healthy.

The obvious thing for Jeon to have done would have been to discard all the diseased amoebas and start again with a fresh, uninfected batch, but he was curious about the nature of the disease that had infected them. Studying the diseased amoebas through a microscope, he found that they contained tens of thousands of tiny, rod-shaped bacteria which had invaded the cells of the amoebas, disrupting their biological processes and killing them. And yet, a few of the 'bacteriarized' amoebas were struggling along, albeit with difficulty. Jeon decided to keep a colony of the survivors, and watch what happened to them.

For five years Jeon tended the colony, regularly picking out the ones that fed and reproduced best and letting the others die. This is a classic example of the kind of artificial selection used by pigeon breeders or dog fanciers. But because of the rapid life cycle of the amoebas, the effects of selection became apparent much more quickly than they do when breeding mammals or birds, which take much longer to reproduce. At the end of the five years, Jeon had a colony of bacteriarized amoebas that seemed healthy in all respects, and were dividing every day, just like their non-infected cousins. Each of these amoebas, though, still contained as many as 40,000 bacteria.

This is still not the end of the story. Usually, with amoebas of the same species, it is possible to tweak out the nuclei from two different cells and swap them over. The amoebas with the swapped nuclei happily carry on growing and dividing, since each nucleus carries essentially the same DNA 'message', and each cell contains essentially the same chemical 'factories'. But when Jeon removed the nuclei from the new strain of bacteriarized amoebas and put them into non-bacteriarized amoebas from the original species whose nuclei had been removed, the cells without the bacteria died. And yet, if, just before it died, the 'clean' cell containing the new nucleus was deliberately infected with a few of the bacteria, they reproduced and increased up to a population of about 40,000, while the cell recovered its health. The nuclei from the bacteriarized cells could no longer function properly without the bacteria. Partly as a result of an accidental infection, and partly thanks to Jeon's artificial selection, a new symbiotic species of amoeba had been created.

This particular new species would probably not have survived without his aid. Even the few original infected cells that did reproduce were very sensitive to heat and cold, and much more easily killed by antibiotics than normal, 'non-bacteriarized' amoebas. But imagine the situation, something over 1.5 billion years ago, when different kinds of cell existed side by side in the waters of the Earth. Large cells, containing a relative wealth of biological material, would have been the prey of smaller bacteria, which would invade them (like the bacteria that invaded Jeon's amoebas) and feast upon the resources they contained. But when the host died, many of the invaders would die as well.

If, however, some of the prey cells evolved a tolerance for their predators, learning to live with colonies of the invaders inside their cells (just like Jeon's amoebas), then this would be good – in the evolutionary sense of ensuring the reproduction of DNA – for the predators as well. They could feed off the leftovers of the biological processes going on in the cell. And (unlike Jeon's amoebas), in this case the invaders seem to have brought a benefit with them, the ability to make more efficient use of energy, giving something back to the invaded cell.

Margulis has argued convincingly that mitochondria got in to what are now complex eukaryotic cells in exactly this way. And the clinching evidence came when researchers found that mitochondria contain their own DNA, quite distinct from the DNA in the nucleus of the cell. This mitochondrial DNA is very similar to the DNA found in some forms of bacteria that live independently. Armed with their own genetic material, mitochondria divide independently of the rest of the cell they live in, and not necessarily at the same time that the cell itself divides. As far as the cell is concerned, all that matters is that there are enough mitochondria to share between the two daughter cells when it does divide.

There is no comparable direct, DNA evidence that the other components of complex eukaryotic cells were incorporated into simpler ancestors from organisms whose ancestors were independent, free-living cells. But it is easy to see how such symbiosis could have arisen, and why it would be evolutionarily beneficial to both the invaded cell and the invader. Since there is direct DNA evidence that mitochondria got into cells in this way, and there is also direct evidence from Jeon's study that bacteria can become symbiotic with amoebas, the case for the symbiotic origins of the cells in our own bodies is as well proven as anything can be when discussing events that happened more than a billion years ago. And this kind of symbiosis does not have to be very common. The trick of symbiosis between a cell and an invading ancestral mitochondrion, for example, may have happened only once, or a handful of times, in the history of the teeming billions of cells in the primeval ocean. As long as the resulting symbionts had an edge over their rivals, they would leave more descendants. With the potential to double their numbers every day (until they reached the limits of food supply or some other resource), it would not take long for the new kind of cell to take its now-prominent place on the evolutionary stage.

Echoing the way in which the cells of our bodies are made up of the descendants of separate, free-living organisms that learned to work together, so our bodies are themselves, of course, made up of thousands of billions of cells. There are about a thousand times as many cells in your body as there are stars in a typical galaxy like our Milky Way. And each of those cells, remember, depends absolutely on the presence of plasmids, doing the work of photosynthesis – even though those plasmids are not in the cells of your body but in the cells of plants, growing on the land surface of the Earth and in the upper layer of the oceans.

Once you start thinking along these lines, it becomes clear that the complex tapestry of life on Earth cannot be unravelled, but has to be taken as a whole. Animals have evolved to make use of (among other things) what plants regard as a waste-product of photosynthesis – excess oxygen, discarded into the atmosphere; plants have evolved to make use of, for example, insects in pollination to help their own reproduction (some species of plant are utterly dependent on insects, and cannot reproduce without their aid); and so on. Lewis Thomas, one of the best writers on biological topics for a general readership, has gone so far as to describe the Earth's atmosphere as 'the world's biggest [cell] membrane'. He meant this not literally, but as a metaphor which highlights the way in which all of life on Earth is interdependent in a way reminiscent of the way in which the four components of a modern eukaryotic cell are interdependent. In his book *Late Night Thoughts on Listening to Mahler's Ninth Symphony*, Thomas wrote that 'Seen from the right distance, from the corner of the eye of an extraterrestrial visitor, [the Earth] must surely seem a single creature.'

But you do not really need to be ET to see the Earth in this way; you just need the right kind of mind. Nevertheless, the extraterrestrial perspective – the view from space – is important. Indeed, it was by thinking about how to find life on Mars that Jim Lovelock was led to conclude, first, that there was no point in sending space probes to look for life on Mars (he pointed out that the information gathered by our instruments on Earth already told us it was dead), and secondly, that any visitor from space would have no difficulty telling that the Earth was alive, long before setting foot (or tentacle) on the surface of our planet.

Lovelock tells how the realization came to him as a flash of insight in 1965, when he was working for NASA, designing instruments that would eventually be used by the Viking Mars probes to sniff the Martian air and look for traces of life products. He saw that there was no need to go to all the trouble and expense of sending a probe to Mars to make these subtle tests, because astronomers already knew that the atmosphere of Mars is inert and must therefore, he reasoned, signify a dead planet.

The atmosphere of Mars contains chemicals bound up in compounds with very low energy, very similar to the exhaust gases from an internal combustion engine. By contrast, the atmosphere of the Earth contains a mixture of gases in a very high energy state. This means that they can very easily react with one another, and with material on the surface of the Earth, to release energy and to produce low-energy compounds. Oxygen in the air, for example, ought, according to simple chemistry, to react vigorously with wood, burning the wood to make carbon dioxide. Oxygen is a high-energy substance, carbon dioxide is a low-energy substance, and, just like water running downhill, simple chemical reactions always seek out the lowest energy.

This is an example of the second law of thermodynamics, regarded as the most fundamental law in all of science. The second law says that things wear out. The Sun, for example, is converting a high-energy form of matter, hydrogen, into a form with lower energy, helium, and pouring the energy thus liberated out into the Universe as it does so. One day, the Sun will have used up all of its nuclear fuel, and will become a cold, burnt-out cinder. This demonstrates the most fundamental version of the second law – heat always flows from a hot object to a cool object, never the other way. Drop an ice-cube into a cup of coffee, and heat flows from the coffee into the ice-cube; the ice melts, and the coffee gets colder. Heat never flows from the ice-cube into the coffee, making the ice even colder and the coffee even hotter.

Human activities, and life processes in general, can reverse this natural order of things on a small scale, but only at the expense of energy stolen from outside. It takes an input of energy to make an ice-cube, for example. Similarly, photosynthesis breaks apart carbon and oxygen in carbon dioxide, and releases the oxygen into the air, but with the aid of solar energy. This is like making water run uphill by using energy from outside – perhaps from an electric pump. But the energy you put in to raise the water will always be less than the energy you get back – perhaps by making the water turn a wheel to run a dynamo – when you let the water run downhill again. Plants get the energy for photosynthesis from sunlight, but the energy released when the carbon and oxygen formed in this way burn again to make carbon dioxide will always be less than the energy used to break the carbon dioxide molecules apart.

The fact that the Earth has an atmosphere rich in oxygen, full of chemical potential energy and highly reactive, is a sign that something out of the ordinary, in chemical terms, is happening on our planet. If the atmosphere of Mars resembles exhaust gases from an internal combustion engine, the atmosphere of the Earth resembles (in fact, in large measure it *is*) the mixture of gases that goes into such an engine. But this is only possible because plants can steal energy from the Sun. Overall, taking the Sun and the planets of the Solar System together, the second law of thermodynamics is not violated, and things are indeed wearing out. This is sometimes described in terms of a quantity called entropy, which *increases* as systems run down towards equilibrium; the entropy of the Solar System as a whole is increasing, as is the entropy of the Universe as a whole.

So a visitor from another star, entering our Solar System, could use a simple spectroscope to investigate the atmospheres of the planets, and conclude that while Venus and Mars, which both have carbon dioxide atmospheres, do not have life, Earth, with its oxygen-rich atmosphere, must have life.

In the mid-1960s, Lovelock's view met with a cool response. If it had been taken seriously, it would have pulled the rug from under the whole Viking project. After all, the main purpose of the project was to look for life on Mars, and Lovelock confidently asserted that there was no life on Mars. He developed the idea quietly (with various colleagues), while continuing his work designing instruments for the mission. By 1968, he had begun to think of the entire ecosystem of the Earth as an interlocking and self-regulating system, and in 1970 his friend and (at the time) neighbour William Golding, the novelist, suggested the name 'Gaia', the Greek goddess of the Earth, for this system. In 1977 the Viking landers confirmed that Mars was indeed as lifeless as

Lovelock had predicted more than ten years previously. By then the idea that life itself controls the environment of the Earth and maintains conditions suitable for life, even in the face of changing outside influences – such as changes in the heat output from the Sun itself – had reached a wide audience, and Lovelock was writing his first book about Gaia.

Today, there are many different views of Gaia, ranging from the semi-religious way in which many people have latched on to the idea of the planet as a living 'Mother Earth', to vehement opposition from some scientists who regard the whole thing as utter nonsense. The truth, like Lovelock's own position on Gaia, undoubtedly lies somewhere between these extremes. It may be stretching things to regard Gaia as a single organism even to the extent that a eukaryotic cell is a single organism, and Lovelock makes no claim that Gaia is in any sense self-aware (let alone intelligent). Yet several key predictions based on the Gaia hypothesis made by Lovelock and his colleagues about the Earth itself have, like the prediction that Mars would be found lifeless, been borne out by subsequent observations. And the Gaia hypothesis seems better able than any other scientific idea to explain a puzzle that has long concerned geologists, astronomers and biologists alike: why was the Earth so warm thousands of millions of years ago when the Sun was young and (relatively) cool and faint?

This is known as the 'faint young Sun paradox'. Astronomers have a very good understanding of the evolution of a star like the Sun. They can test their theoretical models by comparing the predictions from their calculations with observations of similar stars in different stages of their evolution, rather like working out the life cycle of a tree by looking at many trees at different stages of growth in a forest. (Astronomers, somewhat cavalierly, use the word 'evolution' to describe the life cycle of a *single* star; we are not talking about successive generations here, but I doubt that this will cause any confusion.) Since the Sun and the Solar System formed, the brightness of the Sun has increased by about 45 per cent – or, looking at it the other way round, at the time life first appeared on Earth the Sun was about 30 per cent cooler than it is today.

You might think that the puzzle posed by the faint young Sun paradox is why the Earth was not a frozen ball of ice three to four billion years ago, and how there could have been any liquid water around to form the warm little ponds in which evolution began. In fact, the problem is just the reverse. There is no difficulty in explaining how the young Earth managed to stay warm enough for oceans and ponds of liquid water to exist, even though the young Sun was so cool. The Earth formed by the accretion of lumps of rocky material in orbit around the young Sun which collided and stuck together. As the planet formed it was pounded by the impact of many large meteorites which gave up their energy of motion as heat, while the inside of the planet, squeezed by the weight of material accumulating above, became hot enough to melt rock and stimulate volcanic activity. As a result, the gases that surrounded the early Earth must have been rich in compounds like carbon dioxide, released by volcanic activity and by 'outgassing' from the hot rocks of the surface of the newly formed planet. As the meteoritic battering came to an end and the surface of the planet cooled, these gases stopped it from cooling below the freezing point of water, trapping heat through the process known as the greenhouse effect.

It works like this. Most of the Sun's energy is in the form of visible light (this is no surprise, since our eyes have evolved to make use of the radiation that is available), which passes through the atmosphere (even an atmosphere rich in carbon dioxide) without being absorbed. This incoming solar energy warms the surface of the planet, which radiates energy back out towards space in the form of infrared radiation. Some of the infrared radiation, which has longer wavelengths than visible light, is absorbed by gases like carbon dioxide in the atmosphere, which warms the atmosphere near the ground and keeps the whole planet warmer than it would otherwise be. Today, there is just a trace of carbon dioxide in the atmosphere, about 0.03 per cent, but with the aid of water vapour the overall greenhouse effect keeps the Earth about 33°C warmer than it would be if it had no atmosphere. The airless Moon has an average surface temperature of −18°C, while the atmospheric Earth has an average surface temperature of 15°C.

The atmosphere of the young Earth was certainly richer in greenhouse gases than the atmosphere of today, and probably (Lovelock has argued) also thicker (denser) which would also have helped to hold in heat. So it is no surprise that the surface of the young planet was warm enough for liquid water to exist. The intriguing puzzle is that, as the Sun has warmed, the atmosphere has changed its composition in exactly the right way, geological evidence shows, to maintain an almost constant surface temperature. Of course, there have been variations, like the ice ages, but compared with how much the temperature of the Earth would have increased as the Sun warmed had the atmosphere stayed the same, these changes are very small. The Earth has avoided overheating through a 'runaway' greenhouse effect, in which so much heat would have been trapped that all the water in the oceans would have evaporated, turning the planet into a baking, arid desert. Venus has gone through just this process; that planet still has an original, inert carbon dioxide atmosphere, and it is a baking hot, lifeless desert where temperatures can soar above 500°C. But if life had ever got a grip on Venus, perhaps it would now be a living twin to the Earth.

According to the Gaia hypothesis, the action of life itself has ensured that the temperature of the Earth has stayed within the range that life can tolerate. If carbon dioxide were to build up (perhaps as a result of volcanic activity), that would encourage more photosynthesis, which would take the excess carbon dioxide out of the air; if the carbon dioxide content were to fall to the point where plants began to suffer from the cold as the greenhouse effect was reduced, there would be less photosynthesis so more carbon dioxide would stay in the air, and the world would warm.

This is a very simple overview of the possible 'Gaian' way in which life has adjusted the surface temperature of the Earth. Lovelock's argument is rather more subtle, and involves both the removal of carbon dioxide from the air by photosynthesis, and the release of another greenhouse gas, methane, into the air by biological activity. But we do not have to go into all these subtleties, because he has also come up with a very neat analogy to explain how the temperature of a planet can be kept roughly constant by the effect of life processes, even though none of the participating life-forms has any conscious knowledge of what the temperature 'ought' to be. This appealing model of the way a living planet can control its own temperature is called 'Daisyworld'.

Daisyworld is an imaginary planet the same size as the Earth, and the same distance from an imaginary star just like our Sun. The surface of the planet is mainly land, to provide somewhere for the daisies that are the main plant life of the planet to grow (you could imagine the same sort of thing with water-lilies instead of daisies, if you prefer a watery planet). Some of the daisies are light in colour, some are dark, and some are in between; and seeds from a particular shade of daisy always produce flowers the same shade as their parent. If the temperature drops below 5°C, the daisies die of cold; if it rises above 40°C, they die of heat. They thrive at 20°C. To make the calculations simple, Lovelock assumes that the amount of carbon dioxide and other gases in the air of Daisyworld stays the same, and that rain falls only at night, so there are no clouds during the daytime to block out the heat of the sun.

A dark surface absorbs more heat than a light surface, as anyone who has got into a black car that has been parked in direct sunlight will know. Lovelock's Daisyworld model starts when the brightness of the planet's sun has increased just to the point where the temperature on Daisyworld reaches 5°C at the equator. Then, daisy seeds are spread over the planet and it is left to its own devices.

In the first generation of daisies there will be a mixture of shades, all the way from black to white. But as black daisies begin to grow, their dark flowers will absorb incoming solar heat and warm them above the background temperature, so that they will do well. Daisies with white flowers will reflect away the sunlight, cooling the ground in their shade below 5°C. So they will die.

In successive generations, dark daisies are favoured at first, leaving more offspring and spreading across the planet. As they do so, the surface of Daisyworld warms and the whole planet becomes more efficient at absorbing heat from its sun. At the same time, the temperature of the sun is increasing, like that of our own Sun.

But once the temperature rises above 20°C, dark daisies are at a disadvantage. They get too hot for comfort, and although they survive they do less well than white daisies, which are better able to cool off. Now it is the turn of white daisies to reproduce more effectively than black daisies, and to take over territory abandoned by their wilting black counterparts.

For a long time, as the proportion of white daisies increases while the proportion of black daisies decreases, the temperature of Daisyworld stays at around 20°C, even though the temperature of the sun is increasing. There comes a time, however, when all the planet is covered by white daisies, reflecting away as much heat as possible, but the temperature of the Sun is still going up. Eventually the temperature rises above 40°C, and all the daisies die; from then on, the temperature of the planet increases in line with the increasing temperature of the sun.

In answer to criticisms that this Daisyworld model is too simple to take seriously, Lovelock has refined it considerably. He has allowed for twenty different species of daisies, all with slightly different colours, and added in the effects of 'rabbits' that feed on the daisies and 'foxes' that prey on the rabbits. All of this takes place in a computer model which can even simulate what happens when a sudden plague kills off 30 per cent of the daisies. All the details can be found in Lovelock's book *The Ages of Gaia*. The bottom line remains the same – a very simple feedback involving the effect of temperature on living things really can maintain the temperature of a planet like the Earth at the same level for a very long time in the face of steadily increasing energy output from the Sun. If life processes in which carbon dioxide or methane play a part are similarly responsive to temperature changes, it is very easy to understand in principle how the Earth has kept its cool while the Sun's output has increased, even if we do not yet understand all the details of the feedback mechanisms.

But we can get a better idea of the kinds of feedback mechanism that may help the Gaian regulation of temperature on Earth itself by looking at just one detailed example: the interaction of life with both the amount of carbon dioxide in the air and the amount of cloud cover. This is the neatest and most beautiful example of Gaia at work yet elaborated in detail. Tiny creatures called plankton and algae that live in the upper layers of the ocean take up carbon dioxide through photosynthesis. The plankton store some of the carbon as carbonates in their shells, which eventually get deposited in chalky layers on the sea floor; very many marine algae excrete (among other things) quantities of a substance called dimethyl sulphide, or DMS.

Lovelock became interested in the life cycle of these organisms because sulphur is an important element for life, but one which is constantly being lost from the land as sulphates in river water run off to the sea. Somehow, if the concept of Gaia has any meaning at all, there had to be a way in which sulphur was recycled back from the sea to the land; the DMS released to the air by marine algae seemed an ideal vehicle to do the carrying. In the 1980s, researchers found that vast quantities of DMS are released to the air from marine algae, ample to do the job required. But the algae would not, of course, have evolved the ability to excrete DMS simply for the benefit of life on land, so what is in it for them?

Lovelock knew that many marine organisms have difficulty keeping the salt in seawater from penetrating their cells. The sodium chloride in the ocean tends to permeate into the cells through the membrane, and disrupt the life processes going on inside. One way the salt can be kept at bay is by building up a suitable pressure of a non-toxic compound inside the cell, and the compound that many marine algae use for this purpose is built around sulphur, and is known as dimethylsulphonio propionate. Because sulphur is abundant in the oceans (having been washed down off the land), life in the oceans has evolved to make use of it; the release of DMS into the air, recycling the sulphur back to the land, is a by-product of the process.

But once the DMS is in the air, it does not simply float back to the land. Once it had been realized just how much DMS there is in the air over the oceans, Robert Charlson and colleagues at the University of Washington, Seattle, suggested that it might explain another puzzle – how clouds form far from land.

It is very difficult for clouds to grow unless there are tiny particles, known as 'seeds', in the air. The water droplets that build up to make clouds and rain begin their growth on microscopic specks of material in the air, and will not grow without these seeds even if the air is very moist. Over land, and over coastal waters, there are plenty of potential cloud seeds in the form of wind-blown dust, organic material of various kinds, pollution from human activities and the material coughed out from time to time by volcanoes. But how do clouds get started far from land, over mid-ocean? Once DMS gets into the air, it would react with oxygen to produce tiny droplets of sulphuric acid, and Charlson realized that these could provide the seeds on which clouds grow. And, intriguingly, observations of the Pacific Ocean made by weather satellites have shown that clouds tend to form along the routes used regularly by merchant shipping, where seeds that stimulate the formation of clouds are carried into the air in the exhaust from the funnels of ships. What polluting particles from ships' funnels can do, DMS can surely do equally well. So what are the implications for Gaia if DMS released by tiny marine organisms really does alter the cloud cover over the oceans of the world?

Life in the oceans depends on a trade-off between getting enough in the way of raw materials – nutrients – and getting enough sunlight for photosynthesis. Even the animals that feed off the photosynthesizers still depend on sunlight. The sunlight is available only in the top layer of the ocean, while the nutrients tend to sink down into the depths. Frustratingly (if marine algae can get frustrated), there are also nutrients in the air, even some distance from land, in the form of dust particles that carry elements and compounds which are important for the chemistry of life. These dust particles do not have the right properties to act as seeds for cloud formation; the ones that do tend to get rained out of the air before getting very far over the oceans.

In the short term, increasing cloud cover will affect the pattern of winds over the oceans. The spread of cloud blocks out some of the Sun's heat, making some regions of the surface cooler than others, changing convection patterns and encouraging the winds to blow, stirring the top layers of the sea and helping to mix nutrients up from the depths. Clouds generally bring rain, of course, and rain will help to bring particles of dust from the continents, rich in nutrients, down from the air into the water. But if the cloud cover spreads too far it will block out too much sunlight, reducing photosynthesis and, as the pace of biological activity in the upper ocean slows as a result, slowing the release of DMS. The clouds will then thin, photosynthesis will be stimulated, the pace of life will quicken, and more DMS will be released. So it is possible that a Gaian balance very similar to the temperature regulation on Daisyworld could be established through a feedback between biological activity, DMS production and cloud cover.

Where do the plankton come in? The answer lies in the frozen layers of water that make up the Antarctic icecap. Core samples drilled from the Antarctic ice contain layers deposited, year by year, further back in time the deeper you drill. The oldest ice samples recovered in this way were formed from snow that fell before the most recent ice age, over a hundred thousand years ago. And because the layers are in chronological order (rather like tree rings), they can be dated and analysed to find out what was in the snow that fell at different times in the past.

The ice cores turn out to contain traces of another sulphur compound, methanesulphonic acid, or MSA, which is known to be produced from DMS. During the most recent ice age the amount of MSA falling in the snows over Antarctica each year was between two and five times as great as the amount falling today. This must mean that the microscopic life-forms of the upper ocean were more active when the world was cooler – and since there is no reason to think that the latest ice age was special in any way, this is probably the normal pattern of activity during ice ages in general.

Why should this be? The answer comes from the ice cores themselves, which contain two other critical pieces of information. First, analysis of bubbles of air trapped in the ice shows that during an ice age there is much less carbon dioxide in the air than there is today – about 0.02 per cent instead of 0.03 per cent. However, this would not reduce the greenhouse effect sufficiently to explain, by itself, why the world is colder during an ice age. The best explanation of the recurring pattern of ice ages is the changing orbital geometry of the Earth as it moves around the Sun, which alters the balance of heat between the seasons. Sometimes the planet experiences warm summers and cold winters, while at other times summers are relatively cool and winters relatively mild, even though the total amount of heat received by the Earth from the Sun over the course of a year stays the same (apart from the slow increase in the Sun's temperature on a geological timescale) from millennium to millennium.

The key to ice ages, it turns out, is that when northern hemisphere summers are sufficiently cool, snow that falls in winter on the land at high latitudes does not melt, but stays and gradually builds up into great ice sheets; the process is reinforced because shiny white snow and ice fields reflect away incoming solar heat, keeping the planet cool – like the white daisies of Daisyworld. The northern hemisphere summers hold the key because it is only in the north that the polar regions are surrounded by land on which snow can settle without melting; in the southen hemisphere, the polar continent itself is permanently covered in ice, but it is surrounded by ocean. An ice age ends only when the changing astronomical influences bring very hot northern summers, capable of melting the ice (even though, paradoxical as it may seem, that means very cold winters, as well, at the end of an ice age).

Fairly obviously, though, whatever else may be involved in making the world cool into an ice age, the reduction in the amount of atmospheric carbon dioxide will indeed help to keep it cool, through a weakening of the greenhouse effect. How is carbon dioxide removed from the air when the world cools? The ice cores also show that not just MSA but dust particles from around the world fell out of the air over Antarctica in much greater quantities during an ice age. The link between all these discoveries came when John Martin and Steve Fitzwater, of the Moss Landing Marine Laboratories in California, found that ordinary dust blown into the air over all the continents of the world contains iron, which acts as a fertilizer for the microscopic plants that live in the upper layers of the oceans. It may be hard, at first, to think of iron as a fertilizer; but it is much easier to appreciate its importance for living things when you realize that iron is an essential component both of the chlorophyll that plants use in photosynthesis and, for example, of the haemoglobin that carries oxygen around in your blood. Without sufficient iron, planktonic plants in the ocean simply cannot use the available sunlight to convert food in the water into biological material.

Today, the waters around Antarctica and the Arctic Ocean are indeed rich in nutrients, such as phosphates and nitrates, that are not being eaten by plants simply because they cannot get enough iron to make the biochemical molecules they need. When more iron is added to seawater samples from these regions, the marine organisms grow rapidly as they feed on all those nutrients; on a large scale, an influx of iron-bearing dust from the land can produce a bloom of life across the sea.

During an ice age the world is dry. Much of the fresh water on the planet is locked up in ice, and because the temperature is low relatively little water evaporates from the oceans, so there is less rainfall than in the present day. Winds blow the dust from the arid land out over the oceans, where the iron in the dust helps marine organisms to grow. One effect of this is that plankton thrive, absorbing carbon dioxide, turning it into carbonates in their shells, and dropping it onto the sea floor when they die.

Another is that marine algae thrive, increasing the cloud cover of the planet – and clouds, like snow-fields and white daisies, reflect away some of the heat of the Sun. Both processes help to keep the world cool and windy, which seems to be just right for the kinds of marine organism involved in these feedback loops.

Geological evidence shows that the natural state of the world today is to be in an ice age. For the past five million years, the climate of our planet has followed a rhythm in which ice ages of about 100 thousand years are separated by 'interglacial periods' (like present-day conditions) about 15 thousand years long. Other factors, especially changes in the tilt of the Earth, which affect the balance of the seasons can *just* manage to pull the planet out of an ice age from time to time, but the cooling influence always wins out after a few millennia.

It is a sobering thought that, if life really is exercising a Gaian control over the temperature of our planet, reminiscent of Daisyworld, the most important biological feedbacks work not through large animals like ourselves, or even large plants like trees, but the microscopic flora that fill the upper layers of the oceans. It also gives us a rather different perspective from our usual human-centred view of our place among the life on Earth. On this evidence, Gaia would 'prefer' rather cooler conditions than we are used to, conditions which would enable those marine organisms to thrive. And that highlights another problem facing the living planet.

The amount of carbon dioxide left in the air today is tiny indeed – 0.03 per cent of the atmosphere. But the Sun will carry on warming steadily for several billion years yet. Carbon dioxide is the main natural greenhouse gas in the atmosphere, alongside water vapour. Since life cannot stop water from evaporating, drawing down carbon dioxide from the air and depositing it in carbonate sediments and fossil fuel deposits is still the most effective way Gaia has of keeping the planet cool. It may be, if Lovelock is correct, that Gaia is nearing the end of a long phase of stability, rather like Daisyworld warming towards the critical temperature of 40°C. Within a relatively short time, compared with the three billion plus years of the history of life on Earth so far, Gaia may be heading for a convulsion in which the temperature will rise dramatically before (hopefully) some new mechanism evolves to bring things under control again.

Which brings us to our own, human, contributions to all this. Most people are now well aware that the amount of carbon dioxide in the air is increasing rapidly as a result of human activities. By burning coal and oil we put more carbon dioxide into the air, returning into circulation carbon from reservoirs painstakingly laid down by life processes millions of years ago. By destroying forests, we reduce the ability of the biosphere to absorb carbon dioxide. Following the end of the latest ice age, more than ten thousand years ago, the amount of carbon dioxide in the air stayed roughly constant, the ice cores show, at a concentration a little below 0.03 per cent; over the past hundred years, the concentration of carbon dioxide in the air has increased by a quarter, from 0.028 per cent to 0.035 per cent. If this build-up of carbon dioxide continues, the planet will warm as the greenhouse effect gets stronger. How much of this warming can Gaia cope with?

If the concept of Gaia means anything at all, then it is not too fanciful, some people argue, to extend the idea to regard humankind as a disease, infecting the planet and causing an unhealthy rise in temperature that will be detrimental to most other forms of life. Many of the people at the quasi-religious end of the spectrum of opinion about the Gaia hypothesis, though, regard Gaia as a comforting, Mother Earth figure who will 'look after us'. They believe, on the basis of very slender (or non-existent) evidence that somehow natural processes will take the carbon dioxide we are pouring into the air out of circulation, and keep the temperature comfortable for people. But making life comfortable for the invader is not the way to get rid of a virus infecting your body. The fever that accompanies the illness may be unpleasant for a time, but it actually helps to kill off the infection. It is only when the virus is disposed of that the body recovers its normal temperature. 'People sometimes have the attitude that Gaia will look after us,' Lovelock once commented to me, 'but that's wrong. Gaia will look after *herself*. And the best way for her to do that might well be to get rid of us.'

One of the most memorable images of the 1970s (it was actually obtained at the end of the 1960s) is the photograph taken by Apollo astronauts of the blue Earth as a fragile bubble of life rising over the horizon of the dead Moon. That single image, on thousands of posters and illustrating dozens of books, did as much as anything to make people aware that we live on a fragile planet, and that our own activities have to be adjusted to take account of the natural processes that sustain life on Earth. The image also, of course, helped to promote the idea of Gaia – the notion that Earth is the living planet – and it certainly proved Lovelock's point that living planets are easily distinguishable from non-living planets even from far away in space. This has all helped to foster the idea that we are not particularly special, as far as the living planet is concerned. People are not the pinnacle of evolution, but just one species among many, less important (until very recently) to the self-regulatory processes that maintain conditions fit for life than microscopic algae and plankton in the sea. We are important to Gaia now only because we threaten to upset those natural self-regulatory processes.

We are still struggling to come to terms with this, and to find ways in which humankind can become an effective part of Gaia, in the way that mitochondria are effective parts of a cell. But even before we are able to resolve those problems, even larger questions are being posed by new observations of the Universe. What is the place of the living planet, and ourselves, in the Universe at large? Is life an accident, or an inevitable consequence of the laws which determine the way the Universe runs? And where do those laws come from? What is it that makes the Universe the way it is? The surprising answer to those questions now seems to be that evolution has been at work on the Universe itself, and that it is a descendant of a long line of universes.

Nobody would try to explain the existence of a creature as subtle and complex as, say, Jim Lovelock by imagining that he was formed in one step, out of some unique creation event. Rather, we infer that he is the product of billions of years of evolution, and has a long line of ancestors. In the same way, it no longer seems reasonable to suggest that something as subtle and complex as the Universe, with its interplay of physical forces and laws, 'just happened' in a unique creation event. There certainly was, as the COBE data have confirmed, a creation event; but it is no more likely to have been unique than the birth of a baby. But before I can explain how something as subtle and complex as the Universe we see around us could have evolved out of something else, I need to show just how subtle and complex the Universe really is. What, indeed, *is* this Universe that I have been referring to so glibly?

PART THREE

What is the Universe?

CHAPTER SIX

Across the Universe

The Universe is everything that we can see, and interact with, and detect with our instruments. Cosmology is the study of the Universe at large (the cosmos), while astronomy is the study of individual things (or types of thing) within the Universe. It is one of the most astonishing achievements of twentieth-century science that astronomers and cosmologists today are able to describe with confidence the broad outlines of the geography of the Universe, a region of space spanning, in round terms, 15 billion light years. It is perhaps even more remarkable that we now have a clear idea of how the Universe evolved, from its birth in the Big Bang 15 billion years ago to the present day. We also understand, in a general way, how the Universe will age and die. And as if geography and history (and futurology) were not enough, astronomers also have a thorough grasp of what might best be called the zoology of the Universe, the nature of the various 'beasts', such as stars, galaxies and quasars, that inhabit it.

All of this has been achieved in essentially one human lifetime, since about 1920. And yet even books about such dramatic new discoveries as the ripples in time found by COBE usually take the Universe at large for granted, skipping lightly over those billions of light years and all they contain. But it is surely worth trying to put the geography, history and zoology of the Universe in context. That way, we can get a better grasp of just how much has already been achieved by twentieth-century science, and how the new discoveries fit in, not as a bolt from the blue but as the latest (though surely not final) achievement of an ongoing, and very securely founded, investigation. The COBE discoveries were not, after all, pulled magically out of a metaphorical hat by a cosmological conjuror, but represent the culmination (so far) of this century's investigations of the Universe around us.

For most people, though, simply coming to terms with the size of our own planet is difficult enough. The Earth is a roughly spherical ball of rock, with a diameter rather less than 13,000 km (nearly 8000 miles) and a circumference of just over 40,000 km (just under 25,000 miles). The atmosphere on which we depend for our existence is about as thick, compared with the size of the Earth, as the skin of an apple is compared with the size of the apple. The troposphere, the lowest layer of the atmosphere and the one in which weather occurs, extends no more than about 15 km (10 miles) above our heads, while the whole atmosphere fades away into interplanetary space well below an altitude of 1000 km (somewhere below 600 miles).

You can get a feel for what these numbers mean by thinking in terms of how long it would take you to travel by car, or on foot, over the appropriate distances. You could drive the equivalent of the distance from the surface of the Earth to the top of the troposphere in about 15 minutes, even at a modest 60 kph (40 mph); at a brisk walking pace of 6 kph (4 mph), it would take only about two and a half hours. If you could drive a car straight up through the atmosphere, you would be in space in less than a day.

In an aeroplane, you could travel from Europe to Australia, just about the longest journey you can make on Earth, in less than a day, while TV pictures and telephone calls from around the world are bounced into our living rooms with scarcely a discernable delay, at the speed of light – 300,000 km (186,000 miles) a second. This gives us the impression that our planet is an intimate and cosy place – the 'global village' – where, at a price, you can be in one location today and literally anywhere else on Earth tomorrow. But to travel from one side of the Earth to the other by car, at a steady 60 kph, would take nearly two weeks, without allowing for any stops to eat, sleep, attend to other bodily functions or service the vehicle (and without allowing for the difficulty of crossing the oceans).

At the end of the 1980s, Michael Palin re-enacted the journey of Jules Verne's fictitious Phileas Fogg without using air transport. He only just managed to circumnavigate the globe in the required 80 days, averaging around 500 km (310 miles) per day, or 21 kph (13 mph). To *walk* from Europe to Australia, if it were dry land all the way, would take ten times as long as to drive at 60 kph, for the walking alone. If you managed 60 km a day, allowing for sleep and all the rest, it would take more than 300 days, the best part of a year, with no time off for sightseeing en route. And it would take another year to walk home again. So maybe 'global village' is not *quite* the right image, after all.

Yet our planet, nearly 13,000 km across and weighing in at all but 6 thousand billion billion tonnes in mass, is an insignificant dust mote in the Universe at large. The average distance to our nearest neighbour, the Moon, is just over 384,000 km (nearly 239,000 miles), a distance so great that it takes light itself just over one and a quarter seconds to make the journey. Radio waves travel at the speed of light. So if an astronaut on the Moon asks a question of ground control here on Earth, and ground control replies the instant they hear the question, the astronaut has to wait while it takes more than two and a half seconds, from the end of the question asked on the Moon, for the answer to get back.

By terrestrial standards it is a long way to the Moon, but by astronomical standards the Moon and the Earth are such close neighbours that they are best regarded as a single system, a double planet. After all, the Moon is about one-quarter the size of the Earth, and weighs in at about one-eightieth of the Earth's mass. The other planets in our Solar System that have natural satellites typically have moons with masses about one-thousandth that of the 'parent', so our Moon is more than ten times too big to be counted as an ordinary moon.

This may have very deep significance for life on Earth. Because the Moon is so big and (by astronomical standards) so close to the Earth, the tides raised in the oceans by its gravitational influence are correspondingly large. Water moving up and down the shoreline rhythmically, twice every day, has constantly refilled and left stranded pools on the beach and affected the lives of any living creatures within those pools. This may well have played a part in helping life to get started in those warm little ponds; and it surely must have played a part in encouraging life to move out of the oceans, first into the wetlands of the tidal zone and then onto dry land proper.

The simple image of the Moon orbiting around the Earth is not, though, completely accurate. Because the Moon is such a large fraction of the Earth's size it makes more sense to regard the two planets as orbiting around their mutual centre of gravity, like ice-skaters facing each other and holding hands while spinning in a circle. The centre of gravity of the double system itself orbits around the Sun, held by the Sun's gravitational grip, as if, as well as spinning face to face, the ice-skaters were circling around the edge of the rink. Both the Earth and the Moon orbit around the Sun, and their two orbits are interwoven in a dance which takes first the Earth and then the Moon a little closer to the Sun, while the partner always balances this by moving out slightly further away from the Sun.

But now 'slightly' really is the operative word. If you thought the distance from the Earth to the Moon was big, at 384,000 km, compare this with the distance from the Earth–Moon system to the Sun, an impressive 150 million km (around 93 million miles). It takes light from the surface of the Sun, travelling at 300,000 km per second, more than eight minutes to reach the Earth. If the Sun 'went out' instantaneously, it would be more than eight minutes before the sky over the Earth went dark, because at any time there is that much light on its way to us. And the Earth is so far from the Sun that it takes a full year, travelling at an average speed of rather more than 100,000 kph (nearly 30 km per second, 0.01 per cent of the speed of light), to travel once around its orbit, nearly 950 million km in circumference.

The Earth-Moon system is relatively close to the Sun, in the inner part of the Solar System. The Earth is the third planet in the Solar System, counting out from the Sun. It is one of four relatively small, rocky planets – Mercury, Venus, Earth and Mars – that lie within 250 million km of the Sun, and are known as the terrestrial planets. Beyond the orbit of Mars, there is first a jumble of lumps of rock orbiting in a ring known as the asteroid belt; and then, further out, four giant planets, known as the Jovian planets after the largest, Jupiter. Jupiter itself has a volume 1330 times that of the Earth. It is mainly a huge ball of gas, and orbits the Sun at an average distance of 778 million km, taking nearly twelve of our years to orbit the Sun once. Beyond Jupiter there are three other gas planets, Saturn, Uranus and Neptune. Neptune is the farthest known large planet from the Sun, orbiting once every 164.8 years at a distance of roughly 4500 million km.

The last planet, Pluto, is a much smaller ball of ice, not much more than a two-hundredth of the volume of the Earth (less than 30 per cent of the volume of our Moon). Astronomers have recently found another object orbiting the Sun beyond Pluto; it seems not to be a true planet, but a small object, perhaps one of many orbiting together in a belt, rather like the asteroid belt between Mars and Jupiter. Indeed, Pluto itself, judging from its unusual orbit and relatively small size, is probably not a 'real' planet. It may be simply the nearest of those distant asteroids, or possibly a moon that has escaped from the gravitational grip of one of the giant planets. Although the average distance of Pluto from the Sun is 5900 million km, its orbit is very elliptical. Sometimes it is much closer to the Sun than the average, and at other times (in the other half of its orbit) much farther away, so that the actual distance from Pluto to the Sun may be as great as 7375 million km or as 'little' as 4425 million km. It is actually closer to the Sun, for part of its 248 year orbit, than Neptune. Indeed, Pluto is closer to the Sun than Neptune is at present, and will not become the outermost planet in the Solar System again until 1999. At its greatest distance from the Sun, however, Pluto is so remote that light from the Sun takes nearly seven hours, travelling at 300,000 km per second, to reach it. Pluto is, on average, about forty times as far from the Sun as we are. And still our entire Solar System, from the Sun to Pluto, represents just a tiny speck in the cosmic void.

Beyond Pluto – and beyond the putative outer asteroid belt – is the realm of the comets. Those frozen icebergs of space, that may be so important in the origin of life, normally orbit at distances too great for them to be seen even through the most powerful telescopes on Earth. Astronomers believe that the comets are debris left over from the time when the Solar System formed, the cosmic equivalent of rubble. Their orbits are spread around a spherical shell, about 40 *thousand* times as far from the Sun as we are. There may be as many as a hundred billion comets in the cloud; the ones we see are simply the few whose orbits get disturbed and which fall in, past the Jovian planets, to the inner Solar System. There, the heat of the Sun evaporates material from the cometary iceberg, forming a streaming tail that glows in reflected sunlight as the iceberg is whipped around the Sun by gravity and swept back out on its return journey into the cold depths of space, with the tail fading once again from its short-lived glory.

Gravity, the force that swings a comet around the Sun and sends it back out into space, is one of the factors that determines the dimensions of the Solar System. The same force which holds you down on the surface of the Earth and gives you weight holds the Earth and the planets (and the comets) in their orbits around the Sun, and the time it takes for a planet at a certain distance from the Sun to go round the Sun once depends on the strength of the Sun's gravitational pull. Astronomers are able to work out the distances to the planets and the Sun from a combination of surveying techniques and their understanding of the law of gravity, which determines the force that holds each planet in its orbit. And once they know the strength of the Sun's gravitational pull, they can calculate how much mass there is in the Sun.

The surveying techniques are exactly the same ones used here on Earth, but on a larger scale. The key trick is triangulation, which depends on measuring the angle of the line of sight to a distant object from each end of a suitably long base-line.

You can see how this works by holding out a finger and looking at it first with one eye closed and then with the other eye closed. The finger seems to move against the background, because each of your eyes has a slightly different view of it. This apparent shift of a foreground object against the background is called parallax. You could, if you wanted to, measure the distance between your eyes, measure the size of the angle across which your finger seems to move when you switch from one eye to the other, and work out, by drawing a triangle, how far it is from your eyes to your finger. In fact, this is what your brain does automatically, using information from both eyes when judging distances; and that is how you are able, for example, to judge the flight of a ball accurately enough to catch it. Try catching a ball with one eye closed – it is not so easy.

Using longer base-lines and instruments such as theodolites to measure the angles accurately, surveyors can work out the distance to, say, a mountain without ever having to travel to the mountain. With the aid of telescopes on opposite sides of the Earth, so that the base-line is the diameter of our planet, astronomers can measure the distances to the Moon and Mars.

This triangulation trick even gives astronomers the first foothold on the ladder of distances to the stars themselves. By measuring the way in which some stars seem to shift position across the sky when the Earth is on opposite sides of its orbit around the Sun, they can use the diameter of the Earth's orbit, a distance of 300 million km, as the base-line for triangulation. Only the nearest stars are close enough to show any discernible parallax effect, even with such a long base-line. But at least that is a start. The technique gives astronomers a new distance scale, based on parallax, measured in units called parsecs, shortened from 'parallax second of arc'.

Science fiction writers, and sometimes astronomers themselves, talk about distances across space in terms of light years: one light year is the distance travelled by light in one year. (It is important to remember that a light year is a measure of distance, not of time.) A light year is just under 10^{13} km – a distance so vast that nobody can really have a feel for what it means. A parsec (also a measure of distance, not of time) is just over three and a quarter light years, and outside the pages of science fiction you are more likely to find professional astronomers describing distances across our Galaxy in terms of parsecs rather than light years. The nearest star to the Sun, Proxima Centauri, is more than a parsec away, at a distance of 4.2 light years. There are only 31 stars within four parsecs of our Sun.

How can we measure distances to the vast majority of stars, so far away that they show no parallax shift even as the Earth moves from one side of the Sun to the other? The key is that one family of stars, mentioned in Chapter One and known as the Cepheids, has an intriguing and useful property.

All Cepheids vary in brightness, and the nature of the brightness variations depends on the overall brightness of the particular star. The length of the cycle of brightness variations in a Cepheid – the period from one brightness peak to the next – is longer for brighter Cepheids than for fainter ones. The period–luminosity relationship follows a precise mathematical law, a simple equation. So if you know the real brightness of one Cepheid, then the equation tells you how bright any other Cepheid is. And if you know how bright a star really is, then you can work out its distance by measuring how faint it appears to be. Once you know the distance to one Cepheid, you know the distance to them all. Some Cepheids happen to lie in groups of stars held together by gravity, and moving around one another in a cluster. If you know the distance to the Cepheid then you know the distance (more or less) to all the other stars in the cluster as well.

This opens up new possibilities for astronomers, because instead of studying individual stars they can apply statistical techniques to the study of lots of stars. Statistical techniques work best when dealing with large numbers of objects. There are an extraordinary number of stars in our Milky Way Galaxy. If you look at the night sky on the darkest, clearest night that you can ever experience here on Earth, it gleams with what looks at first sight an uncountable array of jewel-like stars. Surprisingly, though, there are only about three thousand stars that can be picked out individually by the unaided human eye at any one time. But on such a clear, dark night you will also see the band of light known as the Milky Way, stretching across the heavens. The Milky Way is the accumulated light of many millions of stars, too faint to be discerned individually except with the aid of telescopes. Telescopic studies show that our Galaxy actually contains, in round terms, about a hundred billion bright stars.

Those stars are spread across a relatively thin disk about 30,000 parsecs (30 kiloparsecs) across, with a central bulge of stars. The relative proportions of the disk and bulge are reminiscent of a fried egg, the yolk representing the bulge and the white the disk. The bulge itself is about 3000 parsecs – 3 kiloparsecs – thick, while the disk is about 800 parsecs thick halfway out from the bulge to its visible edge. The Solar System lies about 10 kiloparsecs out from the centre of the Galaxy, two-thirds of the way to the edge of the visible disk, very much in the Galactic boondocks.

Just as the planets are held in orbit around the Sun by gravity, so the stars in the disk of a galaxy like the Milky Way are held in orbit around the centre of the galaxy by gravity. In this case, though, the gravitational force is the averaged-out influence of all the other stars and stuff in the galaxy, not the effect of a single huge mass at the centre. The Solar System orbits at a speed of 220 km per second, and takes 300 million years to complete one circuit of the Galaxy. Since the Sun was born, it has orbited the Galaxy just 15 times.

The whole system of disk and bulge is surrounded by a spherical halo of stars, a halo with a diameter of about 45 kiloparsecs. But there are relatively few stars in this halo, and most of them are concentrated in about 120 globular clusters. These are systems containing thousands or even millions of individual stars, held together by gravity in a swarm several tens of parsecs across.

But what is a star? Our Sun is the nearest star to Earth, and we know a lot about how stars work in general, and how the Sun works in particular, from studies of many different stars of different sizes and ages, using the ubiquitous techniques of spectroscopy. The life cycle of an individual star may run for billions of years, far too long to be studied by human astronomers.

But a picture of stellar evolution can be gleaned by looking at stars in all stages of the life cycle, just as you could as I have mentioned, investigate the life cycle of a tree by studying a forest for a day, looking at all the different stages in the life of a tree represented by the different trees in the forest.

The bottom line is that a star like our Sun forms out of a cloud of gas and dust in space. It contracts under the pull of its own gravity, and gets hot in the middle as the gas (mainly hydrogen) and dust (perhaps 1 per cent of the mixture) are squeezed together, in exactly the same way that the air in a bicycle pump gets hot when it is squeezed into a tyre. In a star, though, the centre gets so hot that nuclear reaction begin. These reactions convert hydrogen nuclei to helium at a temperature of about 15 million °C, and along the way a small amount of the mass in each hydrogen nucleus is converted into pure energy, in line with Einstein's equation $E = mc^2$.

This is the energy that keeps the Sun and stars shining for billions of years. But it is not, strictly speaking, what makes the Sun hot – gravity does that. Without the nuclear energy holding it up and stopping its contraction, the Sun would be squeezed even harder by gravity, and would get even hotter inside, until something had to give. Paradoxically, nuclear reactions at a temperature of 15 million °C are what keeps the Sun so *cool*! In the process, the Sun itself converts an amount of matter equivalent to about a million elephants into pure energy each second – but it contains so much hydrogen that it has been doing this for nearly five billion years without doing much more than scratch the surface of its fuel reserves, and can do so for another five billion years before running into problems.

The radius of the Sun is a little more than a hundred times the radius of the Earth. Because volume goes as the radius cubed, and 100^3 is 1,000,000, the volume of the Sun is about a million times that of the Earth, or a thousand times that of Jupiter, the largest planet in the Solar System, which weighs in at rather more than 300 Earth masses. The size of the Sun in proportion to Jupiter is more or less the same as the size of Jupiter in proportion to the

Earth. The total mass of the Sun is about 330,000 times that of the Earth. The weight of all this gas pressing down on the core of the Sun not only raises the temperature to 15 million °C, but also produces a pressure three hundred billion times the pressure of the Earth's atmosphere at the surface of our planet. The density of this hot solar core is about twelve times that of solid lead.

These are the conditions under which hydrogen can be transformed into helium. Everyday hydrogen gas, remember, is made up of atoms which each consist of a single proton (the nucleus, which carries most of the mass of the atom and a positive charge), accompanied by an electron (which is much lighter and carries a negative charge). Under the conditions prevailing at the heart of a star, the electrons and protons are separated from one another and mingle freely in a fluid state known as a plasma. This makes the solar core behave just like a gas, even though it is so dense.

Hydrogen is the simplest kind of atom, and the proton is the simplest kind of atomic nucleus. The next most complex stable nucleus is that of helium. It contains two protons, and two neutrons, which are particles almost identical to protons except that they have no electric charge. The four particles in the nucleus of this kind of helium (known, for obvious reasons, as helium-4) are held together by a force known as the strong nuclear force. This force, which holds all nuclei together, has a very short range – just enough to grab hold of the particle next door. But it is so powerful that it can overcome the tendency of the positively charged protons to repel each other, like the repulsion between two magnets if you try to press them together with both north poles (or both south poles) in contact.

The nuclear fusion processes that release energy inside the Sun do so by converting hydrogen nuclei, four at a time, into helium nuclei. The total mass of one helium nucleus is just 0.7 per cent less than the mass of four protons added together, and every time a helium nucleus is made in this way, that much mass is turned into pure energy. It puts the incomprehensibly large mass of the Sun into some sort of perspective to learn that, even converting five million tonnes of mass into energy every second, after five billion years the Sun has so far worked its way through just 4 per cent of its original stock of hydrogen. Only 0.7 per cent of that 4 per cent of its original mass has actually been 'lost' in the form of energy radiated away into space. The conversion of mass into energy is such an efficient way to make energy that all the energy radiated by the Sun in its lifetime to date is equivalent to about a hundred times the mass of the Earth, or rather less than one-third of the mass of Jupiter. The process is so efficient because the mass in Einstein's famous equation is multiplied by c^2: c, the speed of light, is huge, so c^2 is huger still.

When the Sun begins to run into problems with its energy supply, five or six billion years from now, it will not be because it has run out of hydrogen altogether. There will still be plenty of protons in the outer part of the star. But in a star like the Sun, problems arise when it begins to run out of hydrogen in the core, where the temperature and pressure are great enough for nuclear reactions to take place. When this happens, the star's core shrinks and gets hotter still. At a high enough temperature, other fusion reactions can take place, converting helium into carbon, and then into successively more complex nuclei. All of the material here on Earth, and in your body, has been processed in this way inside stars.

There are plenty of stars like the Sun which evolve in this way. The Sun is a typical star, and virtually all the stars you can see in the sky are burning hydrogen into helium quietly in their cores. Such stars are known as members of the 'main sequence', and differ only in mass and brightness. The Sun is an average, ordinary main sequence star, in the middle range of both mass and brightness. After its life on the main sequence, when the core heats up and helium burning begins, the outer parts of a star swell up under the influence of the extra heat within, and the star becomes a red giant. The Sun, like all main sequence stars, will one day become a red giant. Eventually, though, any star will run out of fuel completely. There will be no more nuclear fusion processes that can hold it up against the inward tug of gravity.

When this day comes, most stars simply fade away, cooling into stellar cinders. The fate of our Sun is to cool into a white dwarf star, with just about as much mass as it has today, squeezed into a solid lump about as big as the Earth. But a minority of stars, those that start out with much more mass than our Sun, end their lives violently in huge explosions. These explosions, called supernovae, do two things. First, they blast away the outer layers of the star, spreading the rich mixture of heavy elements built up from hydrogen and helium into space. The dispersed heavy elements mix with clouds of hydrogen and helium gas, forming the material out of which new stars and planets will later condense. We owe our existence to past generations of stars which died violent deaths in this way and laced the interstellar clouds with such essentials (for us) as carbon, oxygen and nitrogen. Secondly, the stellar explosion squeezes the inner core of the star more violently than anything it has experienced before. The remnant, left behind as the debris of the explosion clears, may be a neutron star, in which about as much mass as there is in our Sun is compressed into the volume of a large mountain here on Earth; or it may be that ultimate product of gravity, a completely collapsed object – a black hole. Black holes have a key part to play in the story that will unfold in the final part of the book. But now, having run quickly through the life cycle of a typical star like our Sun, it is time to put the place of the Sun among the stars of the Milky Way in some sort of perspective.

Nobody can really get a grasp of what distances measured in light years or kiloparsecs really mean. But there is a remarkably neat analogy, first pointed out to me by Marcus Chown, which puts cosmic distances in an everyday perspective. I have used the analogy before, but it is so good that that is no reason not to use it again.

If you imagine the Sun reduced to be about the size of a chocolate sweet known as M&Ms or Smarties, then the nearest star would be represented by another chocolate sweet about 150 km (90 miles) away. Apart from stars that belong to binary systems (or more complex aggregations held together by gravity), this is entirely typical of the separation between stars. The distance from one star to its nearest neighbour is several tens of millions of times the diameter of the star itself, and the hundred billion or so stars in a typical galaxy like our own Milky Way are spread over a correspondingly large volume. Galaxies are very big, compared with stars or solar systems.

But the Milky Way is far from being the end of the cosmic story. There are indeed other galaxies out there, beyond the Milky Way. They are so far away that all but the nearest show up only as fuzzy patches of light, even with the aid of the largest telescopes on Earth. Because of this, as we saw in Chapter One, it was only in the 1920s that these 'nebulae' were in fact identified as extragalactic objects, other galaxies in their own right.

Very soon after the identification of these external galaxies, the distances to the nearest ones were measured by the Cepheid technique. Fortunately, one or two galaxies are so close that individual bright stars can be picked out through large telescopes. Then, as mentioned in Chapter One, Edwin Hubble discovered the famous redshift–distance law, which makes it possible to determine the distance to *any* galaxy that we can see.

The redshift is a bodily movement of the characteristic spectral fingerprint of a galaxy (or, in principle, an individual star) from the blue end of the visible spectrum towards the red end. This means that every feature in the spectrum has been shifted to a longer wavelength, in a regular way. The explanation, remember, is that the light from the galaxy has literally been stretched on its journey to us across space; and the reason for that, explained by Albert Einstein's general theory of relativity, is that space itself has stretched during the time the light has been in transit.

This stretching of space is actually predicted by the equations of the general theory (even though Einstein did not believe it at first), and is a key component of the Big Bang theory. Hubble's discovery and Einstein's equations helped to establish the Big Bang model as the best available theory of the origin of the Universe. But apart from these deeper implications, Hubble's law tells us simply that the redshift in the light from a distant galaxy is proportional to its distance from us. In other words, as explained in Chapter One, once you know the distance to one galaxy from Cepheid measurements, and you can measure its redshift, then you can work out the distance to any other galaxy once you have measured its redshift.

At last, on the scale of galaxies, we have reached a level of cosmic size which is relevant to the evolution of the entire Universe. Millions of galaxies have now been photographed and mapped, and it is clear that they form clusters and filamentary structures, tracing out skeins of light across the sky like wisps around great black voids in which very few bright galaxies are seen. We can put these structures into perspective by using a modification of the chocolate-sweet model of the Universe.

Now, imagine that our entire Milky Way Galaxy is the size of the chocolate sweet. On this scale, the nearest large galaxy, known as M31, is another chocolate sweet, not scores of kilometres but just 13 cm (5 inches) away. This is a little misleading, because just as some stars belong to binary (and more complex) systems, so galaxies tend to group together. The Milky Way and M31 are, in fact, held together by gravity, part of an association of galaxies known as the Local Group. But the distance to the nearest equivalent small group of galaxies, known as the Sculptor Group, is still only 60 cm (24 inches) on the chocolate-sweet scale. And just 3 m (10 feet) away we find a huge collection of about two hundred chocolate sweets (each representing a galaxy of about a hundred billion stars) spread over the volume of a basket ball. The next big cluster of galaxies is just 20 m (65 feet) away, and there are more distant clusters which are themselves as much as 20 m across. But the entire visible Universe, out as far as our telescopes can probe, would fit within a sphere just 1 km across, on the scale in which a chocolate sweet represents our Galaxy.

In proportional terms, even the distance to the Virgo Cluster is only six hundred times the diameter of our own Galaxy, while M31 is a mere twenty-five Milky Way diameters from us.

Galaxies are *much* closer together, relatively speaking, than the stars within a galaxy. If galaxies were as far apart, in relative terms, as stars are, then the distance to our nearest galactic neighbour would actually be about a hundred times greater than the most distant object ever seen through our telescopes. Nobody on Earth would ever have seen it, and astronomers would think that the Milky Way was the only galaxy in the Universe. Extragalactic space is far richer in galaxies than galaxies are in stars, and that is why cosmologists are able to glean an understanding of the way in which matter is distributed across the Universe, and how that distribution of matter has changed as the Universe has evolved.

There is another reason why we seem to be able to understand the Universe at large. It is a very simple place. The behaviour of the whole Universe depends on two main factors – how fast it is expanding, and how much matter there is in it. The matter, through its gravitational influence, tends to hold back the expansion, so we have two simple processes at work, opposing (or balancing) each other. The equations that describe the behaviour of the Universe are a little more complex than those that describe the balance of a see-saw. But with only two opposing effects to worry about it is still quite easy to predict what will happen next, and the only really interesting question is which influence will win the tug-of-war.

The behaviour of living organisms like ourselves is much more complicated. There are usually more than two possible outcomes when human beings interact, and no sociologist or psychologist can describe human behaviour in terms of equations as simple as those of the general theory of relativity. Of course, there are simpler life-forms that do behave in much more predictable ways; but none of the components of the Universe at large exhibits the complexity of human behaviour. Stars, for example, are very simple because they contain atomic matter stripped to its basics – protons, neutrons and electrons interacting in a plasma. People are made of complicated molecules, such as DNA. Each cell in your body contains enough coiled up DNA to stretch over a distance of 180 cm (70 inches). Think of this in terms of a helix untwisted to make a simple ladder – there are 2.9 billion rungs on this ladder, with a complex hydrocarbon molecule attached to each rung. The more complex the molecular structure, the harder it is to understand its behaviour. Extreme conditions, especially heat, break things down into their simplest component parts, and under those conditions the laws of physics do a good job of explaining, or predicting, what goes on. The Big Bang was even hotter than the inside of a star, and in that sense even simpler, which is why the laws of physics are so good at describing the Universe at large. It is, in fact, no surprise that we understand the Universe better than we understand ourselves.

There is another way of relating the complexity of a human being to the complexity – or lack of complexity – of cosmic systems. The basic unit of your own body is the cell, and the basic unit of the Milky Way Galaxy is a star. There are about a thousand times as many cells in your body as there are stars in our Galaxy. Leaving aside, for now, the question of whether the Galaxy itself can be said to be alive in any meaningful sense of the term, a galaxy is clearly a simpler 'entity' than a human being. And although there are many millions of galaxies in the Universe, they are arranged in simpler patterns than the stars of a galaxy like our own. The farther out we look, the simpler things seem to be, until what COBE reveals is an almost uniform sea of microwave radiation laced with just the merest hints of structure, those famous ripples.

Like the background radiation itself, light from distant galaxies is not just light from far away in space; it is also light from far away in time, because it takes millions of years for light to cross intergalactic space, even at the speed of 300,000 km per second. The most distant objects known have redshifts so great that the light we see them by must have started on its journey when the Universe was just a tenth of its present age, within 2 billion years of the Big Bang itself. These objects are the quasars mentioned earlier. The only reason we can see them across such vast distances is that they pour out energy in the most profligate fashion, radiating as much energy as the entire hundred billion stars of our own Galaxy from a volume no bigger across than our Solar System.

The only way in which such energy can be produced is if each quasar is a black hole, containing about a hundred million times as much mass as our Sun (or 0.1 per cent of the mass of a galaxy), lying at the heart of a young galaxy. As material falls in towards such a black hole, it is accelerated by gravity to enormous velocities, and forms a hot, swirling ring of material around the black hole. Black holes are the ultimate manifestation of the power of gravity: nothing at all – not even light – can escape from them, and that is how they get their name. But under such extreme conditions, a sizeable fraction of the mass in the infalling material is turned into pure energy and radiated away before the matter falls into the hole and is lost forever to the outside world.

Black holes like this probably lie at the hearts of virtually all galaxies, including our Milky Way. But as the galaxies age and settle down, there is less and less spare matter around to 'feed' the central black hole. So the violent activity typical of quasars is also typical of the early Universe, and there is no activity like this going on in the Local Group today. Nevertheless, even these 'dead' black holes are vitally important to the story of the Universe (and black holes are discussed in more detail in the next chapter). First, though, I want to look at another aspect of gravity at work that comes in to the story of the Universe at large.

The ubiquitous redshift can, as we saw in Chapter Two, do more than just tell us the distance to a galaxy or quasar. In a swarm of galaxies like the Virgo Cluster, all roughly the same distance from us, it is fairly straightforward to pick out the galaxies that are physically part of the cluster. They can be distinguished from any galaxies with much higher or lower redshifts that might lie in the same direction on the sky but which are much farther away or closer to us than the cluster itself. But there is still a spread of redshifts for the galaxies in the cluster proper, a distribution around the average value. It is like measuring the height of a hundred or so children all the same age and taking the average; not every child has average height for their age, and there is a distribution of heights around the average. Not every galaxy in such a cluster has precisely the same average redshift, because as well as the redshift caused by the stretching of space their spectra are affected by their motion through space.

A galaxy moving away from us through space also shows an extra redshift, proportional to its velocity, while a galaxy actually moving towards us through space tries to show the opposite effect in its spectrum, a blueshift. It's like trying to walk down an upward-moving escalator. You still get carried upward, but not so quickly as if you stood still and let the escalator carry you along. The moving escalator is like expanding space, carrying galaxies along; the person trying to walk down it is like a galaxy moving towards us through space, even while being carried away from us by the expansion of space. Indeed, the galaxy M31 is moving towards us through space, and because it is such a near neighbour this overwhelms the cosmological redshift effect (which, remember, is proportional to distance, and so is very small for near neighbours). The light from M31 really is blueshifted as seen from Earth.

With more distant galaxies, like those in the Virgo Cluster and beyond, the net effect is always to produce a redshift, because at those distances the cosmological effect is so big it overwhelms any smaller blueshift produced by the motion of a galaxy through space. But the individual motions of individual galaxies through space either add to the cosmological effect, making the individual redshift bigger, or subtract from it, making the individual redshift smaller. As a result, astronomers can work out how fast the galaxies in a cluster are moving relative to one another by comparing individual galaxy redshifts with the average for their cluster. It is like measuring the heights of those children not in absolute terms, but by how much taller or shorter each child is than the average for the group. It turns out, as I mentioned earlier, that in every case the speeds at which the galaxies in a group are moving relative to one another are so great that the clusters cannot possibly be held together by the gravity of all the bright stars in all the galaxies in the cluster. Since the galaxies are indeed being held tightly in a cluster, and since gravity is the only force known that could do the holding together, this must mean that there is a lot more dark matter within the clusters, holding them together.

The same sort of thing happens within a galaxy like our own. The redshift and blueshift effect can be used to look at the way in which a galaxy rotates. In every case, the outer parts of the galaxy are found to be rotating too fast to be held in place by the gravity of all the visible stars in the galaxy. Our Solar System's leisurely once-in-200-million-years orbit of the Milky Way may seem sober enough by terrestrial standards, but even our Galaxy must be held together by the gravitational influence of a vast halo of dark matter, in which all the bright stuff is embedded.

The observations show that individual galaxies may contain ten times as much dark stuff as bright stuff, while clusters and superclusters of galaxies are held together by the gravitational influence of as much as a hundred times more dark matter than the bright matter we can see. Everything in the visible Universe – everything ever studied with the aid of telescopes here on Earth – makes up only about 1 per cent of the material content of the cosmos.

It is not just that there must be a lot of dark stuff in the form of lumps of rock, or clouds of cold gas and dust in space. By using Einstein's equations to 'wind back' the expansion of the Universe to the Big Bang, cosmologists can calculate the conditions of temperature and pressure in which the Universe was born. Those are the calculations that explain how the material from which stars are made was processed in the Big Bang, producing the mixture of 75 per cent hydrogen and 25 per cent helium out of which the first stars were made. It is not just that the heavy elements of which we and planet Earth are made represent barely 1 per cent of the mass in bright stars; the bright stars themselves represent less than 1 per cent of the stuff of the

Universe. Everything else must be in a quite different form of matter, nothing like everyday atoms. Nobody knows for sure what this dark stuff is, but it is definitely not ordinary protons, neutrons and electrons. And the pattern of the ripples in the cosmic background radiation found by COBE is exactly in line with the kind of structure that should have been imprinted on the Universe, according to standard cosmological theories, by such dark matter.

There is a very profound implication in all of this. Just as clusters of galaxies are bound together by gravity, then so is the whole Universe – provided all the dark matter hinted at by galactic motions and COBE's discoveries really is there. Although the theorists already believed that it is there, COBE's observations come as welcome confirmation of those theories. It means that there is so much matter in the Universe that the gravity of everything it contains is strong enough to one day halt and reverse the cosmic expansion into a cosmic collapse. The space between galaxies will shrink, instead of stretching, and light from distant galaxies will be blueshifted, not redshifted, Eventually, everything will be compressed back into a superdense fireball, a mirror-image of the Big Bang, collapsing back towards a singularity.

What happens then is the story I shall tell in Part Four of this book, the story of the living Universe. It means that we are living *inside* a black hole – a black hole which contains literally everything in the Universe: all the stars, galaxies, quasars and dark matter. I mean this literally. The important point is that spacetime is bent around the Universe to make it a closed, self-contained entity. The entire Universe is like a bubble in spacetime. But there are some very peculiar features about that bubble, and about our presence inside it. These strange coincidences have to be explained by any satisfactory theory of cosmology; but before we can attempt that, we have to understand just what those coincidences are, and how black holes themselves work.

CHAPTER SEVEN

The Goldilocks Effect

One of the most surprising things about our Universe is that it is just about as big as it could be without being infinite. This is expressed in terms of Einstein's description of the curvature of spacetime, by saying that the Universe is very nearly 'flat'. And this in turn seems to be the result of an extraordinary coincidence, back at the birth of the Universe in the Big Bang, which balanced the inward tug of gravity against the outward expansion of the Universe with exquisite precision. This may seem like a rather unusual use of the term 'balance'. After all, the Universe expanded out of the Big Bang with dramatic speed, in spite of the best efforts of gravity to prevent it. But the balance can be seen in terms of what happens later in the life of the Universe. Gravity is *exactly* strong enough to *just* cancel out the expansion and bring it to a halt. Not so strong that the expansion never got started, and not so weak that the expansion will never end. It is rather as if you could throw a baseball upwards from the foot of a skyscraper at exactly the right speed so that it would halt its upward flight precisely when it was level with the top of the building, so that a friend on the observation deck could reach out and grab the ball while it seemed, to her, to be hovering in the air. The Universe will one day 'hover' like this, at a state of maximum expansion, before very reluctantly beginning to fall back upon itself. As we shall soon see, this really does require a very delicate balancing of gravity and expansion.

I am going to argue that all apparent coincidences of cosmology are neither extraordinary nor accidental, but are the natural result of the processes that gave birth to the Universe. First, though, I want to look at some of those coincidences, and see just what it is that any satisfactory theory of the birth of the Universe must explain.

I mentioned in Chapter One that, according to the equations of the general theory of relativity, space may be gently curved, bending round on itself so that the entire Universe is closed. In that picture, the three-dimensional Universe would be a self-contained entity in the same way that the two-dimensional surface of the Earth is a self-contained entity, curved back on itself to make the surface of a sphere. One of the key features of the surface of a sphere is that it has a definite area – it is finite. In the same way, if the Universe is closed round on itself like this, but in an extra dimension, it must have a finite volume in three dimensions.

In fact, the general theory allows for two other possibilities. The opposite of a closed universe is, logically enough, an 'open' universe. In terms of geometry, just as you can think of a closed universe as the three-dimensional equivalent of the surface of a sphere, so you can think of an open universe as the three-dimensional equivalent of a saddle-shaped surface, the shape of a mountain pass. The important thing about this kind of surface is that (unlike a real saddle) it stretches away to infinity, without ever coming to an end. If the Universe were open, then it too would stretch away to infinity in all directions.

The third possibility allowed by Einstein's equations is that of a 'flat' universe. A flat universe is the three-dimensional equivalent of a flat piece of paper. It also stretches away to infinity, but there is no spacetime curvature in a flat universe, which is a special case of the equations, exactly balanced between being open and closed.

As far as the expanding Universe is concerned, there is another way to look at the difference between open and closed – one which relates directly to the way in which the Universe has expanded out of the Big Bang, and continues to expand today. There are two opposed processes at work in the Universe on the largest scale. The outward expansion, triggered by the processes of the Big Bang itself, is opposed by the pull of gravity, trying to halt the expansion. Ever since the Big Bang, the speed at which the Universe expands has been slowing down. Will gravity ever overwhelm the expansion, first bringing it to a halt and then making the entire Universe collapse back upon itself? Or will the expansion continue forever, albeit at a slower and slower rate?

In the first case, the Universe must be closed, in the Einsteinian sense of the word. The second case corresponds to an open universe. And in a flat universe the expansion slows and slows as the aeons pass, until gravity *just* brings it to a halt, without ever managing to reverse the expansion into a collapse.

Imagine throwing a ball straight up from the surface of the Earth. If you throw it with normal human strength, it will go up a certain distance and then fall back. If you had superhuman strength, you could throw the ball so fast that it would escape from the pull of the Earth's gravity and travel out across space forever. And if you had not only superhuman strength but also superhuman skill, you could throw the ball at exactly the right speed so that gravity would eventually bring it very, very nearly to a halt, but it would never quite stop and fall back to Earth.

From the surface of the Earth, it is quite easy to escape into space. Rockets launching satellites and space probes do it often. But if the Earth had more mass, and therefore a stronger gravitational pull, it would be correspondingly harder to escape. You need a less powerful rocket (or ball-throwing arm) to escape from the surface of the Moon, because the Moon is smaller and less massive than the Earth; but you would need a more powerful rocket to escape from Jupiter, because Jupiter is more massive than the Earth.

If the gravity at the surface of an object is so strong that nothing at all can escape, not even light, then the object is called a black hole. If you were inside a black hole and tried to shine a torch beam upwards, even the light itself would be bent back, by the curvature of space, towards your feet. If the black hole were very big (if the entire Universe *is* a black hole) the light beam might travel for billions of years on its journey out into space, around the Universe and back to your feet, but the end-result would still be the same. The spacetime around a black hole is completely closed upon itself by gravity.

This is, indeed, exactly equivalent to the closed Universe. When we describe the Universe as closed, we mean that the gravity of all the matter in the Universe put together is so strong that nothing, not even light, can ever escape its grip. Space is closed upon itself, and one day gravity will overcome the expansion of the Universe and make everything fall together into a single point of superdensity, a singularity. The Universe is, literally, a black hole. The reason why it does not fit our popular image of a black hole is simply that it is so big. Anyone who knows anything about black holes knows that all the matter and energy inside is destined to fall into the centre of the black hole and be crushed out of existence. But nothing can travel faster than light, so in a black hole a billion light years across (say), this crushing process will take a billion years. Our Universe is a great deal bigger than a billion light years across, and all the stuff it contains is actually expanding *out* from a singularity (do not worry about where the singularity came from in the first place; all will soon be made clear). It will take many billions of years before everything gets crushed back into the singularity and things return to the state we think of as normal for a black hole – 'what once was, and one day again will be'. The Universe today is a black hole, and so was the cosmic fireball, in spite of their apparent differences, in the same way that both an adult person and a newly born baby are human beings, in spite of their apparent differences.

But the Universe is only *just* a black hole – gravity only just balances the outward expansion. What decides this is the density of matter in the Universe at any cosmic epoch. The strength of gravity depends on the density of matter, not just on the amount of matter. A neutron star, for example, might contain the same amount of matter as our Sun; but the strength of gravity at the surface of the neutron star will be much stronger, because that matter is packed into a smaller volume, and therefore has a higher density. When the Universe was younger and denser, its gravitational pull, tugging inward upon itself, was stronger – but when the Universe was younger, it was also expanding faster than it is today. Obviously, the matter gets spread thinner as the Universe expands, but there is still the same amount of matter there. As the matter thins and the gravitational pull gets weaker, the expansion rate also slows, so less gravitational effort is needed to keep the Universe closed. The way the two effects change together ensures that if gravity were ever strong enough to do the trick of making the Universe closed, it will always be enough, no matter how much the Universe expands and thins.

At any epoch during the life of the Universe there is a critical density, corresponding to flatness in the Einsteinian sense. If the density exceeds the critical value, the Universe is closed; a density less than the critical value corresponds to an open Universe. A closed Universe is the variation on Einstein's theme that can be represented by the skin of an expanding balloon. By saying that the Universe is very nearly flat, we are saying that the skin of the balloon is only curved very, very gently – the Universe is like the skin of an extremely large balloon.

By adding up the mass of all the bright stars in all the galaxies in our neighbourhood, astronomers can account for about 1 per cent of the critical density. Adding in the influence of all the dark matter that is revealed by the way galaxies move within clusters and superclusters raises this to at least 10 per cent of the critical density, probably more. Observations of the way galaxies move are not accurate enough to tell us exactly how much matter there is altogether, but they do also show that there is definitely no more than ten times the critical density of matter present in the Universe today. Fifteen billion years after the Big Bang, the density of the Universe is balanced between one-tenth and ten times the critical value corresponding to flatness. If you set the critical density to be 1, then the actual density of the Universe today is somewhere between 0.1 and 10.

In fact, with the COBE observations, inflation and the theory of galaxy formation to guide us, we can pin the parameter down even more tightly. Inflation, remember, is the process that whooshed up the Universe from a seed smaller than the nucleus of an atom to something very much larger indeed, all in a split-second of time. Among other things, this answers the puzzle of how the Universe came to be flat. Obviously, the 'balloon skin' corresponding to the original tiny seed would be very tightly curved, and nowhere near flat; it is inflation that gives the initial burst of expansion that increases the size of the balloon ('inflates it') so much that it then looks nearly flat. Imagine inflating a child's balloon to the size of the Earth without bursting it, and you begin to have some idea of what went on. The surface of that balloon would, like the surface of the Earth, seem flat to anyone sitting on it. But the relative inflation that took place when the Universe was born involved far greater expansion than that.

Combining inflation theory with the COBE data and our understanding of galaxy formation, we know that the actual density of the Universe must be at least 1, when all the dark matter is included; but it still cannot be more than 10. Pinning down the range of possibilities to within a factor of 10 may not sound so remarkable – even politicians usually manage to get their budget forecasts more accurate than that. But the point is that the critical density is the only 'special' density in Einstein's equations. As far as we know, there is nothing built into the laws of physics to say that the density could not have had any value at all – a billion times the critical density, or one billionth of the critical density, or any other number you might pluck out of the air. If the Universe could have any density at all, why should it just happen, by coincidence, to sit right on the dividing line between being open and being closed – why should it be as big as possible without being infinite?

This discovery is even more surprising because, throughout the past 15 billion years, the density must have been getting further away from the critical value. It is very difficult to balance the opposing mechanisms of expansion and gravity, and as time passes an expanding universe deviates more and more from flatness. It is as if you were to fold a piece of stiff cardboard along its centre to make an inverted 'V', then tried to balance a marble on the crest of the resulting ridge. The marble would quickly roll off on one side or the other, and as it rolled it would get further and further away from the ridge. It would take superhuman skill to balance the marble so skilfully that it stayed on top of the ridge for 15 billion years – but the Universe seems to have been set up, in the beginning, with just such a precise balance between being open and being closed.

That ridge, the centre-line, corresponds to perfect universal flatness, in between the two alternatives of being open or closed. We know from the COBE results and the theory of inflation that the Universe must, in fact, be closed, and always has been closed. If the critical density is set at 1, then the density of the real Universe must be more than 1, even if only just fractionally so. If the Universe is closed to start with, gravity holds back the expansion and keeps the density high, which holds back the expansion still more, and so on. If it had been open to start with, the expansion would have thinned out matter so that gravity gets weaker, and this would allow more thinning out, and so on. If the Universe is nearly flat today, it must have been just a little flatter last year, or last millennium; and it must have been *much* closer to perfect flatness long ago when it was young.

How much closer? In order to be within a factor of ten of perfect flatness today, one second after the moment of its birth the Universe must have been flat to one part in 10^{15}: the amount by which the density differed from the critical value was represented by a decimal point followed by 14 zeros and a 1. But we know that the density must always have been more than the critical value. If the value of the critical density is set at 1, then when the Universe was one second old the actual density was between 1 and 1.000 000 000 000 001.

How much further back can we push the calculation? It may seem presumptuous to talk blithely of what was happening when the Universe was just one second old, 15 billion years ago, but astronomers are actually quite confident that they understand what was going on at that time. It is fairly simple to 'wind back' the equations of Einstein's theory to discover the temperature and pressure of the Universe as far back as the first second or so, and these were no more extreme than the most extreme conditions found in our Solar System today – the conditions at the heart of the Sun. Then, the entire Universe was in a hot fireball state rather like the inside of the Sun today, and the behaviour of particles like protons and neutrons under such conditions is well understood. Earlier epochs in the life of the Universe, when its age was measured in split seconds, can be described by slightly less well-understood theories, requiring more exotic particles and the application of quantum theory. This is the basis for the highly successful idea of inflation, which explains how a very smooth Universe containing enough dark matter to make spacetime closed (just!) and to begin to pull galaxies together to make those ripples observed by COBE, emerged from the Big Bang.

But even quantum theory has its limitations, and no scientific theory can properly describe the singularity at the moment of creation itself. A singularity is a place (a point in spacetime) where the laws of physics break down completely. Whatever material form an object might take – a star, a human being, a broken washing-machine or whatever – its identity is completely lost as it is crushed out of existence in the singularity. The very atoms and particles of which a thing is made are destroyed and crushed out of existence.

Roger Penrose proved in the 1960s that every black hole must contain a singularity. The ultimate fate of everything inside a black hole is to form one point of infinite density, where all bets are off and anything can happen. Stephen Hawking took up this idea and proved that the expanding Universe must have emerged *out of* a singularity. Just as anything falling into a singularity loses its identity and merges into the singularity itself, so energy emerging from a singularity can take on any identity. Anything at all can emerge from a singularity. The laws of physics do not apply to the split second in which energy or mass emerges from a singularity, and in principle a singularity could emit broken washing-machines, human beings or stars. It just happens that 'our' singularity emitted a very nearly flat fireball of mass–energy.

But one law does still apply. Simple things are more likely to emerge than complex things, and the simplest thing of all is energy itself. So it is no surprise that the structure of the very first split second of the birth of the Universe was energy filling a tiny, closed region of space and sending it blasting outward in the expansion that became the Big Bang. That first split second represents the limit of quantum theory. It corresponds to 10^{-43} of a second. This ridiculously small number is the 'quantum' of time, the smallest time that has any physical meaning. As far as the laws of physics that operate in the Universe are concerned, the Universe was 'born' with an age of 10^{-43} of a second: there was no earlier time.

If you take the rather extreme extrapolation of the flatness of the Universe right back to this moment, known as the Planck time (after Max Planck, one of the pioneers of quantum physics), it means that a Universe that is flat to one part in ten today, and was flat to one part in 10^{15} at the age of one second, must have been flat to one part in 10^{60} at the moment of creation, the Planck time. This means that the 'flatness parameter' is the most accurately determined number in the whole of science. If it really were just a coincidence that the Universe happened to be born out of a singularity in this state, that really would be remarkable. It is even more remarkable when you consider how even a tiny deviation from flatness during the first second of the Universe would have quickly been magnified, making life impossible.

At the moment the Universe was born (the Planck time), its diameter was just over 10^{-33} cm – the distance that light could travel in the Planck time. This is known as the Planck length. All the mass–energy of the Universe (exactly as much as is contained today in all the bright stars and galaxies and all the dark matter put together) was squeezed into this tiny distance, creating an intense gravitational field that was almost precisely balanced by the almost exactly equally intense outward expansion of this seed of creation.

But these conditions are so extreme that they make the mind boggle. Let us move forward in our imagination a second or so, to the time when the Universe was only as hot and dense, everywhere, as the inside of a star. Suppose that at the age of one second the density of the Universe had been just ten times the critical density. In that case, gravity would have been so strong that the Universe would have collapsed back into a singularity almost immediately, without ever giving a chance for galaxies, stars, planets and people to form. Or suppose that at the age of one second the Universe had only one-tenth of the critical density. Gravity would then have been so feeble that, even with the aid of dark matter, clouds of gas would never have been pulled together to make galaxies and stars and planets and people. The runaway Universe would have expanded rapidly, spreading the leftover matter from the Big Bang ever more thinly across the void. The flatness parameter really does have to have been set as accurately as one part in 10^{15} when the Universe was just one second old for us to be here at all. Was it just a lucky coincidence that the Universe emerged from the Big Bang in this state, just right to ensure our ultimate arrival on the scene?

This is an example of what is known as anthropic cosmology. Anthropic cosmology is an attempt to find a reason for the relationship between human beings and the cosmos. You could also call it the Goldilocks principle, since, like Baby Bear's porridge, the Universe seems to be 'just right' for us. The Goldilocks principle says that coincidences like this cannot really be coincidences at all, but must have some deep significance. At one extreme there are those who would argue that the Universe seems to have been designed for our benefit – that it was deliberately set up so that, for example, the flatness parameter was just right to ensure that stars and planets and people could evolve. Another point of view is that it is indeed no more than an extraordinary coincidence, pure luck, that the singularity out of which our Universe emerged just happened to set things up with the flatness parameter balanced to this ultrafine degree between instant recollapse and rapid runaway. But the most intriguing possibility is that there is a reason why the outburst from the singularity should have produced a Universe that is *just* closed, expanding as fast as it possibly can while still being a black hole. That reason may have nothing to do with the presence of people in the Universe today; it may indeed be a lucky accident that we are here, because the conditions that have naturally evolved in the Universe *for other reasons* just happen to favour us. Nevertheless, the extent to which those coincidences of cosmology do favour us is truly astonishing.

The power of anthropic reasoning in providing insight into the workings of the Universe is superbly demonstrated by an argument that raged in the nineteenth century. The protaganists were the evolutionary biologists, headed by Charles Darwin, and the physicists and astronomers, headed by William Thomson (later Lord Kelvin).

Darwin's theory of evolution could explain how all living things on Earth had descended from a simple common ancestor, as a result of competition and natural selection leading to the survival of the fittest. But above all, evolution clearly required an enormous span of time in which to do its work. Geological processes also required very long periods of time in which to shape the landscape by processes such as erosion. Darwin and the geologists concluded that the Earth must have been around for an immeasurably long time, at least a billion years, in more or less the same state that it is in today.

But the astronomers of the day knew of no process which could have kept the Sun shining for so long. If the Sun were made entirely of coal, for example, burning in an atmosphere of pure oxygen, it would be burnt out in a few tens of thousands of years. The most powerful energy source that Thomson could imagine was gravity. He realized that if the Sun were slowly shrinking inward under its own weight, then gravitational energy would gradually be converted into heat, which could be radiated away into space. But a straightforward calculation showed that even if the Sun shrank from a ball of gas more than a hundred times the diameter of the Earth to a solid lump about as big as the

Earth itself, shining all the while as brightly as it does now, all the gravitational energy would be released in a few tens of millions of years. Its maximum possible lifetime, according to the laws of physics known to Thomson, was certainly less than a hundred million years, and far less than anything required by geology and evolution.

And yet, we exist. That fact alone is sufficient, with hindsight, to pinpoint the error being made by Thomson. No known form of energy, in the nineteenth century, could keep the Sun hot for long enough for evolution to have done its work here on Earth, so some unknown form of energy must be at work inside the Sun. This argument was put forward forcefully in 1899 by the American astronomer Thomas Chamberlain. Within a few years, Einstein's special theory of relativity predicted that heat could be released by the conversion of mass into energy; within a few decades, astrophysicists had a detailed understanding of exactly how this process operates in the Sun, converting hydrogen nuclei into helium nuclei and releasing a little mass–energy in the process. The power of the Goldilocks principle is that it can tell us when our treasured theories are wrong (or incomplete) even when we do not yet know how to put them right (or make them more complete). In effect, Darwin's theory of evolution 'predicted' Einstein's theory of relativity.

Such anthropic predictions do not always require the benefit of hindsight. I have already mentioned that Fred Hoyle was the prime mover, in the 1950s, behind the work that led to an understanding of how all the elements except for hydrogen, helium and tiny traces of very light elements left over from the Big Bang are manufactured inside stars. This includes the carbon, nitrogen and oxygen that are so fundamentally important (along with that hydrogen) to life on Earth. But that understanding rests upon a powerful piece of insight by Hoyle that led to a genuine prediction, based on the fact that we exist, of what the properties of subatomic particles must be.

Once physicists began to understand how nuclear reactions work, it was fairly straightforward to guess, in a general sort of way, how the heavy elements are built up inside stars. The first step is the combination of hydrogen nuclei (single protons) into nuclei of helium-4, which each contain two protons and two neutrons. This is the process going on inside the Sun and other main sequence stars. The 4 in helium-4 simply denotes the total number of nucleons (that is, protons plus neutrons) in each nucleus of the element, and is very important.

Helium-4, as it happens, is a very stable nucleus, and behaves in many ways like a single particle – which is why it is often called the alpha particle. This stability makes nuclei that contain a whole number of helium-4 nuclei (a whole number of alpha particles) particularly stable and common in the Universe. They include carbon-12 and oxygen-16, which can be regarded as being made up of three and four alpha particles, respectively. Once carbon-12 is present inside a star, then under suitable conditions of temperature and pressure it is relatively easy to add an extra helium-4 nucleus to make oxygen-16, and so on in fours all the way up the nuclear fusion ladder to iron-56.

In-between elements (like nitrogen-14) are made when lighter neighbours (in this case carbon-12) latch onto the odd proton or two from the surrounding superdense plasma, or when radioactive forms of some nuclei decay, spitting out the odd proton or positron (the positively charged counterpart of an electron) and readjusting their nuclear composition. Still heavier elements, beyond iron-56, are made in supernova explosions, which come in to the story of the life of a galaxy, told in the next chapter.

But there is a snag. The exception to the rule that nuclei made up from alpha particles are particularly stable breaks down drastically in one crucial place – on the very first rung up the ladder of nuclear fusion. The nucleus composed of two alpha particles put together is beryllium-8; and beryllium-8 is so unstable that any nucleus of beryllium-8 that happens to form when two helium-4 nuclei collide in the plasma maelstrom inside a star blasts itself apart within 10^{-19} a second. The only way to make carbon-12 (and therefore the only way to make anything heavier than carbon-12) seemed to be if a third alpha particle arrived on the scene during that tiny split second in which a beryllium-8 nucleus existed. But under such circumstances kinetic energy provided by the impact of a third alpha particle would itself simply blow the beryllium-8 nucleus apart.

In 1954, Hoyle saw that there was only one way round the difficulty. It hinges on a property known as resonance. Atomic nuclei, including those of carbon-12, can exist in more than one state. The different states are known as energy levels, and you can think of them as like the different notes corresponding to a single plucked guitar string. The string can vibrate at its fundamental wavelength, one wave filling the length of the string, or it can produce overtone vibrations, harmonics in which the wavelength is one-half, or one-third, or some other integer fraction of the length of the string, so that there are always a whole number of waves filling the length of the string.

If you shout at a guitar, the strings will move a little in response to the waves in the air corresponding to the noise – but only a little, because the wavelengths of the sound you make do not correspond to natural wavelengths for the guitar strings. But if you play a note on another instrument at a wavelength that corresponds exactly to one of the harmonics of the guitar strings, the string will vibrate in sympathy, as if the guitar is playing itself.

Nuclei 'resonate' in an analogous way. If the right amount of energy is put into a carbon-12 nucleus, it will absorb the energy and move into an excited state for a while, before radiating the energy away again and falling back to its lowest energy level, known as the ground state. Hoyle said that the only way in which carbon-12 could be formed from a collision between highly unstable beryllium-8 and an alpha particle, during the split second that the beryllium-8 nucleus exists, would be if carbon-12 had a suitable excited state corresponding to the energy of one beryllium-8 nucleus and one alpha particle put together.

It is as if the 'shout' of an alpha particle hitting a beryllium-8 nucleus just happens to be in tune with a carbon-12 'harmonic note'. If so, then instead of the collision blowing everything apart, the combined nuclei would slip smoothly into the guise of an excited carbon-12 nucleus. That carbon-12 nucleus could then radiate away its excess energy, and settle down into its ground state.

In 1954, the idea seemed outrageous. In order to make carbon (and all the heavier elements), the laws of physics would have to be precisely fine-tuned for this resonance to occur. Hardly anyone took the idea seriously, and the researchers who did, under Hoyle's nagging, set out to measure the resonances of carbon-12 in the laboratory, did so more in the hope of shutting him up than in expectation of proving him right. But to their surprise, they found exactly what Hoyle had predicted. The coincidence is so remarkable that it is worth putting some numbers in. The energies involved are measured in mega-electronvolts (MeV), but that doesn't matter – just look at the numbers themselves.

The energy of a beryllium-8 nucleus plus a helium-4 nucleus is 7.3667 MeV; the energy of an excited carbon-12 nucleus is 7.6549 MeV. The difference between the two is just under 4 per cent, and the extra 0.3 MeV or so required to make the match perfect is just the sort of energy of motion that will be carried into the collision by the alpha particle. Instead of blasting the beryllium-8 apart, the energy carried by the alpha particle is just enough to carry the combination over the top and make excited carbon-12: 'just right' to make the carbon nuclei that go into the atoms in Baby Bear's porridge.

The power of the coincidence is brought home when you consider what would have happened if the balance had just tilted the other way, with the beryllium-8/alpha particle combination having an energy even just 1 per cent more than the excited state of carbon-12. Then there would be no resonance. The incoming alpha particle would indeed blast the beryllium-8 apart, and there would be no carbon (or any heavier elements) in the Universe. And all for the sake of a swing of just 5 per cent on the excited carbon-12 energy level.

This is not the end of the story. Other nuclei can also have excited states. Oxygen-16, for example, has an energy level at

7.1187 MeV, but the combined energy of a carbon-12 nucleus and an alpha particle is 7.1616 MeV. This is just too high (by a mere 0.6 per cent) for a combination of carbon-12 and helium-4 to yield a perfect match with oxygen-16. In this case the kinetic energy of the incoming alpha particle, adding to the total, makes the discrepancy even worse.

There is no problem about making oxygen-16 slowly inside stars, because carbon-12 (unlike beryllium-8) is stable and stays around to get involved in many collisions. For a proportion of carbon-12 nuclei, nuclear reactions resulting from these collisions eventually do the trick, without recourse to resonance. But plenty of carbon is left over at the end of the life of the star, ready to form part of the mix going into the next generation of stars and planets. If, however, the oxygen-16 energy level lay just above the combined carbon-12/alpha particle level (which would happen if the excited oxygen energy level were just 1 per cent higher), then it would resonate, and all the carbon-12 inside a star would quickly be converted into oxygen-16. There would be no carbon available to take part in all those interesting reactions that are the basis of life as we know it on Earth.

So there are two remarkable coincidences which allow life-forms like us to exist – a double Goldilocks effect – at work inside stars. The first makes it possible for carbon to form at all; the second prevents all the carbon turning into oxygen, but allows enough oxygen to be made to carry the process of nucleosynthesis on up the chain to iron-56. The fact that we exist, with bodies based on carbon chemistry and breathing oxygen from the air, tells us in very precise terms what some of the properties of nuclear particles must be. Could such coincidences arise by accident? Were they 'built in' to the structure of the Universe by a designer? Or have they evolved through natural selection?

Hoyle's own view was that if the Universe is infinite then there might be regions of the Universe in which different laws of physics operate. In some regions (most!) the coincidences that allow carbon and oxygen to form inside stars would not exist, and there would be no life-forms like us to notice the fact. But in some parts of the Universe where the laws of physics just happened to allow this (literally by coincidence), then life-forms like us could evolve to notice the fact and wonder at its implications. In Hoyle's picture, it is as if Goldilocks had an infinite number of bowls of porridge to choose from, with temperatures ranging all the way from boiling hot to stone cold, and with all possible degrees of saltiness for each temperature. One of the bowls must contain porridge that is 'just right' for her taste.

The idea that the Universe is open and infinite in this sense has been overtaken by recent events, although as we shall see in Chapter Nine a different perspective on infinity may come into the debate. But I find this version of anthropic reasoning very unsatisfying, since it holds out no hope of ever finding why things are as they are. The entire Universe, including ourselves, is explained as no more than a lucky accident (or series of lucky accidents). Since there is no prospect of ever finding why the laws of physics are as they are (because in this picture there is no 'why'), physicists might just as well shut up shop and turn their attention to something else. Fortunately, though, the new understanding of the Big Bang does suggest a 'why'.

The range of coincidences that have to be explained by any satisfactory theory of how and why the Universe came to be as it is is vast. If a coincidence in the energy levels of atomic nuclei inside stars seems a little remote from everyday life, there are plenty of examples closer to home. Take the peculiar properties of water, for example. In the second decade of the twentieth century, Lawrence Henderson of Harvard University drew attention to the remarkable properties of water that seem to be essential for the existence of life as we know it. The ability of water to dissolve such a wide range of other substances is unique, and it has other properties that distinguish it from other liquids. I shall pick on just one – the fact that water expands on freezing into ice, so that ice floats on water.

This is an utterly bizarre property. You would not expect, for example, a lump of solid lead to float in a puddle of liquid lead, and you would be right. The strange behaviour of ice only seems 'normal' to us because there is a lot of water in our environment. It is also very important for life on Earth. If ice sank when it froze, the way solids 'ought' to when freezing out from the liquid form of the same substance, then in the winter ice would settle at the bottom of lakes and oceans, not the top. The exposed surface of the water would continue to lose energy, radiating it away, and more ice would literally 'build up' from the bottom until the entire ocean was one lump of ice. In such a frozen condition, with the shiny white surface reflecting away solar heat, it would be extremely difficult ever to thaw the planet again. But in the real world, ice forms on the surface of the sea and stops evaporation, which helps to keep the water below warm. It also acts as an insulating lid, stopping heat from being radiated away. Ice is a very good insulator, which is why an igloo made of ice-blocks is a cosy and practical kind of dwelling in the Arctic.

This behaviour can be traced to the unusual properties of hydrogen atoms and oxygen atoms, which together form molecules of water (H_2O). One of the key factors is that because each hydrogen atom only has one electron, the proton inside (the hydrogen nucleus) is not screened very well from the outside world. Part of its positive charge is 'visible' through the electron screen, giving it the ability to form weak bonds with electrons in the clouds around other atoms even when it is already chemically bonded in a molecule like water. This type of bond, known as a hydrogen bond, operates particularly strongly between hydrogen nuclei in water molecules and oxygen atoms in adjacent water molecules. This makes water molecules slightly sticky, so that they do not move past one another as easily as they otherwise would. As a result, water is a liquid at temperatures as high as 100°C, even though the nearest comparable compound, hydrogen sulphide (H_2S), boils below -50°C and is a gas at room temperature.

The hydrogen bonding between molecules is also responsible for the crystalline structure of ice, in which the molecules are arranged in a very open lattice, making the ice lighter than liquid water. And as if that were not enough, among its many other influences on chemistry it is the hydrogen bond that holds the two strands of molecules of DNA together.

The properties of carbon, hydrogen and oxygen seem to be not only essential for the appearance of life on Earth, but unusually 'arranged' to make life possible, in the same way that the properties of carbon-12 and oxygen-16 nuclei, together with those of beryllium-8 and alpha particles, seem to be 'arranged' to make nucleosynthesis possible inside stars (and in the same way that Baby Bear's porridge seems to be 'arranged' to suit Goldilocks' taste).

But while Henderson appreciated the importance of these properties of carbon, hydrogen and oxygen to the evolution of life, he stopped short of suggesting that those properties of the elements had themselves evolved. 'Nothing is more certain', he wrote in his book *The Order of Nature* (published in 1917), 'than that the properties of hydrogen, carbon, and oxygen are changeless throughout time and space.'

Most present-day scientists would agree with this, since they would also agree that the properties of these elements depend on the nature of fundamental forces that operate throughout the Universe, and have done so in exactly the same way since the Big Bang – since, in fact, the moment when the age of the Universe was the same as the Planck time. Indeed, in their book *The Anthropic Cosmological Principle*, published in 1986, two modern researchers, John Barrow and Frank Tipler, quote Henderson with approval, going on to say that 'the elements cannot "evolve" in the sense of having the freedom to take different evolutionary paths like living creatures can. This portion of Henderson's argument must still be regarded as sound.' Later in that same book, they say that:

> twentieth-century physics has discovered there exist invariant properties of the natural world and its elementary components which render inevitable the gross size and structure of almost all its composite objects. The size of bodies like stars, planets and even people are neither random

nor the result of any progressive selection process, but simply manifestations of the different strengths of the various forces of Nature.

I disagree. Not with the detail of the argument used by Henderson and by Barrow and Tipler, but with the narrowness of their vision. The natural forces to which they refer almost certainly are the same throughout the Universe as we know it, so that the properties of stars or planets or water are, beyond any reasonable doubt, going to be the same in any galaxy that we can see through our telescopes as they are in our Milky Way Galaxy. But where did those forces, and the visible Universe itself, come from? What went on before the Big Bang, and is there any reason why the set of forces that emerged from the singularity at the Planck time took on the precise values they did? I believe that all these questions can now be answered, putting the entire Universe in an evolutionary perspective. But before I go on to describe those answers I should explain what these fundamental forces are. They involve just four kinds of interaction between particles, and they underpin every one of the coincidences of cosmology, from the properties of water to the carbon-12 resonance and from the size of a human being to the shape of a galaxy.

Two of the forces are familiar from everyday life. The first is gravity, responsible for holding us down on the surface of the Earth, holding the Earth in orbit around the Sun, and, indeed, for holding the whole Universe together and ensuring that it will one day recollapse into a singularity. The second familiar force is really two forces in one, known as electromagnetism. Until the nineteenth century, the forces of electricity and magnetism seemed to be two separate entities, but they were unified into one package by the equations developed by James Clerk Maxwell. A flowing electric current always has a magnetic field associated with it, and a moving magnetic field induces an electric current in a conductor. This is the basis of both the generation of electricity in power stations and the operation of electric motors. Light and radio waves are both forms of electromagnetic radiation, and the speed at which they travel, the famous constant *c*, comes out of Maxwell's equations of electromagnetism.

Gravity and electromagnetism are both long-range forces. Gravity acts literally right across the Universe, and so too in a sense does electromagnetism, when light travels between widely separated galaxies. But, unlike gravity, both electricity and magnetism come in two flavours which cancel each other out. An atom, a planet like the Earth or a galaxy like the Milky Way contains the same amount of positive electric charge (in the form of protons) as it does negative electric charge (in the form of electrons), so its overall charge is zero. Similarly, every magnetic north pole is accompanied by a magnetic south pole. So the effect of electromagnetism on the Universe at large is much less (or at least, less obvious) than it might otherwise be. Although the orbit of Mars, for example, is slightly affected by the gravity of the Earth (and vice versa), because each planet has no overall electric charge and no overall magnetic polarity, neither orbit is affected directly by electromagnetic forces from the other planet.

The other two forces of nature both operate only on the small scale, over a distance roughly equivalent to the diameter of the nucleus of an atom. A neutron and a proton inside the nucleus, essentially touching each other, can 'feel' the influence of these nuclear forces from each other, but they cannot feel the influence of the nuclear forces from the atom next door, let alone one on a different planet. The strong nuclear force is the powerful (though short-range) force of attraction which holds the nucleus together against the tendency of the positive charge on all the protons to blow the nucleus apart.

The weak force is harder to picture in everyday terms, because it is not simply involved in pulling things together (like gravity) or pushing them apart (like two magnets placed with their north poles in contact). The weak force is responsible for radioactive decay, and in particular for the way in which a neutron can eject an electron and transform itself into a proton.

The behaviour of everything in the visible Universe, and the Universe itself, depends on the balance between these four forces, together with the exact values of the basic unit of electric charge (the charge on an electron) and the exact masses of the proton, neutron and electron. The standard idea of the Big Bang as a unique event offers no explanation of why these fundamental entities should have the properties they do. It is either just luck (coincidence) or the work of a designer, depending on your personal preference. But there is no doubt that the exact values of these fundamental entities are crucially important to the nature of the Universe, and to the existence of life-forms like us.

The four fundamental forces differ wildly not only in range but in strength, and these differences are a key to the evolution of the Universe and to our existence. Their strengths are usually measured in terms of how two protons alongside each other in a nucleus would 'feel' each of the forces. Because electromagnetism was the first of the forces to be studied in the laboratory, the strength of the electromagnetic interaction is set at 1. On this scale, the strength of gravity is a feeble 10^{-38}: the force of electric repulsion between two protons in a nucleus is 10^{38} times stronger than the force of gravity attracting them to each other.

On the same scale, the strength of the weak force is 10^{-10}. Even the weak force is 10^{28} times stronger than gravity, so gravity cannot stop the decay of a neutron into a proton and an electron. But the strength of the strong force is roughly 1000, in these units, comfortably enough to hold the nucleus together even though electric forces are trying to break it apart. It cannot prevent radioactive decay, however, because the electron ejected in the process is immune to the strong force, just as a piece of wood is immune to magnetic forces.

Just as Maxwell succeeded in combining the description of electricity and magnetism into one set of equations, so today physicists dream of finding a single set of equations that will describe all four of these interactions in one package. They have already succeeded in developing an 'electroweak' theory that unifies electromagnetism and the weak force, and they think they have a good idea of how the strong force might be included in the package. Gravity is proving more stubborn. Physicists also believe that the properties of protons and neutrons depend on the fact that each of these particles is made up of three other entities, known as quarks. But none of that need bother us now. We already have enough information to explain why planets, stars and people exist in our Universe.

The size of atoms, for example, can be explained simply in terms of the masses of protons and electrons and the strength of the electric force of attraction between them. Although quantum physics can be used to make the argument more subtle, there is enough truth in the simple picture of an electron 'in orbit' around the nucleus, like a planet orbiting the Sun, for it to be relevant. The balance between the electric force pulling the electron in and the energy of its orbital motion depends on the mass of the electron and the strength of the electric force, and a very simple calculation then sets the sizes of atoms to be a bit less than 10^{-8} cm – the right answer, since atoms really are about that size.

When it comes to planets, what matters is the balance between the electric forces between electrons in the outer parts of atoms and gravity. Gravity tends to pull everything together, but is a very feeble force on the scale of an atom. The electric charge on the electrons in an atom tends to keep them apart (once again, quantum effects are important if you want to do the calculation accurately, but not important enough to bother us here). The outward force resisting the crushing of an atom stays the same for any atom, but the inward force of gravity trying to crush the atom depends on how many atoms there are piled up on top of it. Once you start dealing with 10^{38} atoms or more, you have a chance of gravity overwhelming the electric forces and crushing the atoms in the middle of the pile completely.

In fact, it is not quite that easy, because all those 10^{38} atoms cannot all be at exactly the same point. Instead, they are spread out through a rock, a planet or a star (a rock a few kilometres across actually contains about 10^{40} atoms). This gives gravity a handicap, reducing its effectiveness by a one-third power, because the volume of a lump of material depends on the cube of its radius. So gravity can crush atoms at the centre of a planet out of existence once there are about 10^{57} atoms collected in one planet (38 is two-thirds of 57).

So gravity dominates the electrical force when the total mass of a collection of atoms is about 10^{57} times the mass of a proton (a number I always remember as the 'Heinz soup parameter'). Under these conditions, atoms are crushed out of existence and converted into a plasma of free electrons and nuclei. Nuclear fusion begins, as the nuclei are squeezed closer together and start bouncing off one another. The collection of atoms has become a star. A lump of matter containing 10^{57} atoms is, indeed, the size of a star just a little bit smaller than our Sun (with a mass about 85 per cent of the Sun's).

The balance between gravity and electric forces is also important to life on Earth. It is the electrical forces *between* atoms that holds molecules together, and on the surface of the Earth these forces are rather closely balanced by the strength of the force of gravity from all the atoms in the Earth pulling together.

The old adage 'the bigger they come, the harder they fall' comes into its own here. The electric forces that hold together the atoms of, say, a coffee-cup are likely to be disrupted if I drop the cup onto a hard floor. This is because the cup picks up kinetic energy as it falls under the influence of gravity. When it hits the ground this energy of motion has to go somewhere, and it is shared out among the atoms and molecules of the cup, disrupting the chemical (that is, electrical) bonds that hold them together.

If I drop the cup from just above the floor, it will probably not break. The kinetic energy shared out among the atoms and molecules when it hits the ground will be enough to raise the temperature of the cup by a minute amount, but not enough to break the bonds. The cup is more likely to break if I drop it from a greater height, because it will have had more chance to accelerate under the influence of gravity, and it will hit the ground harder.

It is the same with people, and other land animals. If you break a bone when you fall over, it is because the energy provided by gravity during your fall has been sufficient to break apart the electrical bonds between atoms and molecules in the bone. The only numbers that really matter are the relative strengths of the electromagnetic and gravitational forces, the masses of protons and neutrons, and the number of atoms in your body and in the entire Earth put together. Once again, a simple balancing of these numbers tells us that any animal that lives on Earth and has a mass bigger than about 100 kg (220 lb) has a fair chance of breaking a bone when it falls over. You can be a bit heavier than 100 kg if you are agile enough to avoid falls, and much heavier if you move carefully, like an elephant. But you cannot get away with being big and agile. Which is one reason why there are no 10 metre (30 feet) tall giant people around on Earth.

These are all interesting discoveries, and help to explain why the Universe is as it is, and why we are as we are. But they do not come into the category of cosmic coincidences. The size of a human being, for example, simply represents an adaptation to the forces that operate on the surface of our planet. But when we look again at what goes on inside stars, we find further evidence that there is something very peculiar about the balance of the four forces in our Universe.

With two basic kinds of nucleus (the proton and the neutron) to play with, you might expect that three kinds of 'double nucleon' nucleus could exist: the diproton (two protons stuck together), the dineutron (two neutrons stuck together) and the deuteron (one of each kind stuck together). In fact, only the deuteron can exist in our Universe, and this has profound implications for energy generation inside stars and for nucleosynthesis.

The reason why the diproton and the dineutron are unstable is that they have too much energy. Nothing is ever completely still, and the jostling motion of a particle like a proton is known as its zero-point energy. Two particles (or more) in a nucleus are constantly jostling one another as a result, and this zero-point energy tends to break the nucleus apart, adding to any tendency for disruption caused by the positive charge on all the protons.

There is a further complication introduced by quantum effects, which makes it easier for two different kinds of particle to stick together – as in the deuteron – than for two identical particles to stick together.

The overall effect is that in a deuteron the strong force can *just* hold the two particles together; in a diproton or in a dineutron the combination of electric repulsion and zero-point energy would be too great for the strong force to overcome. This is crucially important. If the diproton did exist, it would be a form of helium – helium-2. If the strong force were just a little bit (13 per cent) stronger than it actually is, all of the hydrogen in the Universe would have been turned into helium-2 very early on in the Big Bang. There would be no hydrogen in the Universe today at all, and therefore no water, and none of the curious coincidences that Henderson noted about the behaviour of water, and no organic life-forms like ourselves. Since all main sequence stars are powered by the fusion of protons into helium-4 nuclei, there would be no stars like the Sun in such a universe, either. On the other hand, if the strong force were rather less strong (by about 31 per cent), even the deuteron could not be held together. The Universe would then consist of nothing but hydrogen, since nucleosynthesis would never get started. Either way, we would not exist.

This is a powerful example of the Goldilocks effect at work. Is the strength of the strong force 'just right' for Goldilocks because, as Hoyle argued, it is part of an infinite variety of choices, one of which must, by chance, be just right? Or has the cosmic porridge been lovingly cooked up by a cosmic chef who knows what Goldilocks likes? Or is there some reason why a Universe like ours must evolve into more or less the state in which we find it?

Whichever alternative you prefer, there is plenty more to explain. Our existence also depends, for example, on the fact that the mass of the neutron is very nearly the same as the mass of the proton. There is no explanation for this in terms of standard physics; it is just one of the facts of life that a neutron weighs just 1.29 MeV more than a proton (since mass is equivalent to energy, both are measured in the same units). It is a puzzle why, first, there should be any difference at all in the masses and, secondly, that the difference should have this particular value. But whatever the reasons for its existence, the value of the mass difference comes into the story of nucleosynthesis and energy generation inside stars because it affects the stability of nuclei.

Again, quantum effects come into play, and I shall not go into the details. What matters is that the so-called binding energy of a nucleus has to be sufficient to overcome the mass difference between the protons and the neutrons. The binding energy in deuterium is just 2.23 MeV. So if the mass difference between a proton and a neutron were twice as big, deuterium could not exist.

While we are considering might-have-beens, imagine what the Universe would be like if the force of gravity were not so weak. Suppose it was about 10 billion times stronger (which would still make it by far the weakest of the four forces, with a strength 10^{-28} that of the electric force). You would not need so many atoms to make a star. The atoms and nuclei themselves would be much the same as in our Universe, because gravity would still be vastly weaker than the other forces and would not affect subatomic processes. But stars in a high-gravity universe would then have the mass of a small planet, or a large asteroid in our Universe; and because they would contain fewer atoms, squeezed more fiercely by gravity, they would use up their nuclear fuel more quickly. Quite a lot more quickly, in fact – in about a year.

A typical star in the high-gravity universe would be about 2 km across, burning with a bright blue light; any planet about 500,000 km from the star (roughly twice the distance from the Earth to the Moon) would be just comfortably warm, with an average temperature of about 25°C. It would also be no bigger than a small moon in our Solar System. But it would be very unlikely that life could evolve on such a planet, in the short lifetime of the parent star. Indeed, at the time nucleosynthesis had produced heavy elements and planets had formed, the entire universe would be only a few years old, and a few light years across.

On the other hand, if gravity were even weaker than it actually is, it would never have been able to hold clouds of gas in an expanding universe together to make galaxies, stars and planets at all. Such a universe would be filled with a thin gas of hydrogen and helium, spreading ever thinner as the universe expanded.

Changing any of the forces of nature, or the fundamental particle masses, seems to shift the balance of our Universe against life. In the same way, if Baby Bear's porridge is 'just right' for Goldilocks, then any change in the recipe would have made the porridge less palatable for her. But at least Goldilocks did have three bowls of porridge to choose from, improving her chances of finding one that she liked. We only have one Universe, take it or leave it.

The more you look at the way our Universe is set up, the more it seems to be set up in a very odd way, to encourage nucleosynthesis and the formation of stars, planets and people. In his book *Galaxies, Nuclei and Quasars* (published in 1965), Hoyle says that it looks as if 'the laws of physics have been deliberately designed with regard to the consequences they produce inside stars'. He has also described the Universe, and the carbon and oxygen resonances in particular, as a 'put-up job'. Quoted by Paul Davies in his book *The Accidental Universe* (published in 1982), Hoyle said that 'a commonsense interpretation of the facts suggests that a superintellect has monkeyed with physics, as well as chemistry and biology, and that there are no blind forces worth speaking about in nature'.

But this is almost exactly the 'argument from design' used, before evolution was understood, as evidence for the existence of God (it still is used, as an alleged refutation of the ideas of evolution, by people who do not – or will not – understand Darwin's theory). The argument was presented in its classic form by William Paley, in the eighteenth century. He said that the fact that a bee, or an oak tree or a human being is so precisely fitted to its environment reveals the work of a great designer, and that such things could not arise by chance. The refutation of this argument is that evolution by natural selection does not operate by chance in the way that the argument from design suggests as the only alternative to the hand of God. Variations from one generation to the next – mutations – may indeed occur at random, but the *selection* process, essential to evolution, is not random at all, and involves selection *of the fittest*.

The argument is beautifully explained by Richard Dawkins, in his book *The Blind Watchmaker*. It works to explain such amazing 'coincidences' among life on Earth as the way in which, say, a humming bird has evolved a beak which can penetrate the flowers of the plants it feeds on. The 'fit' of the bird's beak to the flower is not like the fit of a piece in a jigsaw puzzle, carefully carved out by a designer, but rather like the fit of a flowing river to the river bed, a result of the river and the bed 'evolving' together to match each other. And I believe that evolution can also be used to explain the balance of the four forces of nature, the marginal stability of deuterium, and all the other coincidences of cosmology puzzled over by Hoyle and others.

But I also believe that these puzzles have, until very recently, been looked at from the wrong perspective. It is natural for human beings like us to see the coincidences of cosmology as indicating that the Universe has been set up (either by a designer or by evolution) for our benefit. But this anthropocentric view may be very wide of the mark. Take the example of our own eyes. They have evolved to make very efficient use of the radiation provided by sunlight. Without sunlight (or its artificial equivalent), our eyes would be useless. There is no doubt that artificial lights, like the one in the corner of the room where I wrote this book, were designed for our benefit, with human eyes in mind. But there is no reason at all to think that the Sun was designed to produce light suitable for human eyes. Rather, the light came first and eyes evolved to make use of it.

Indeed, our entire living planet, Gaia, has evolved to make use of energy pouring out of the Sun, in what is essentially a parasitic manner.

Recall Jim Lovelock's comment that 'Gaia will do what's best for herself, not necessarily what's best for us'. The Universe may have evolved along lines that are good for the Universe (or for universes in general), with any benefits for us merely an incidental side-effect. We (by which I mean all organic life) may simply have been very good at making the best use of what the Universe had to offer, just as our eyes have evolved to be very good at making use of what the Sun has to offer. The use of carbon, hydrogen, oxygen and nitrogen provided by the Universe is an equally parasitic act by life as we know it.

In that case, the interesting question is why the production of carbon, hydrogen, oxygen and nitrogen is 'good' for the Universe itself. Why is there nucleosynthesis, and why are there main sequence stars, and galaxies like the Milky Way? What is it that is being selected for out there in the Universe at large?

These are the questions I shall answer in the rest of this book. It may very well be that there is something special about the laws of physics, and that they have been selected by a process of evolution. But there is no evidence at all that that reason involves our existence. It seems more likely to me that the processes of natural selection which have led to the existence of our Universe are completely blind to the existence of organic life-forms like ourselves, and living planets like the Earth. Our existence is an example of the supreme opportunism of life, latching on to any available source of energy and insinuating itself into any conceivable ecological niche. Those niches, we have already seen, extend from the scale of viruses at least up to the scale of our whole planet. Now, we can see how the scale extends indefinitely upwards; and *then*, at last, we shall be able to understand the real driving force of the evolution of universes, and the possibly genuine coincidence that what is good for the Universe just happens to be good for us.

PART FOUR

Is the Universe Alive?

CHAPTER EIGHT

Is the Galaxy Alive?

It may seem outrageous to claim that our Galaxy, a collection of stars, gas and dust (plus the still mysterious dark matter), should be alive. Even those astronomers who use words like 'evolution' to describe the structure of an individual galaxy and the nature of galaxies in general change as the Universe ages do not claim that this is anything other than an analogy, or a metaphor. But I believe that it really is more than an analogy, and that we have been misled into thinking of objects like galaxies as merely inanimate collections of matter by the accidents of scale – the distance scale, and the timescale.

Galaxies, as we have seen, are huge. The Milky Way is tens of thousands of light years across. Our view of it, from within, is rather like that of a microbe trying to assess the structure of a human body, and to come to terms with the fact that this giant entity is a living being. The timescales of the evolution of galaxies and the Universe are even more mind-boggling. Our Galaxy rotates once every few hundred million years, and any dynamic processes of change and development in the structure of the Galaxy are on similar, or longer, timescales.

In a human lifetime, the Galaxy (and the Universe) seems to be static and unchanging, and it is hard for us to unravel the nature of the changes going on in the Universe at large, where a period of billions of years corresponds to merely the first flush of youth of the Universe. We have been able to come to terms with the nature of change in the Universe only because light travels at a finite speed, so that we see different galaxies at different distances from us as they were at different times in the history of the

Universe. It also helps that there are millions of galaxies to study, so that we can get an understanding of their ecology by looking at many different galaxies at different stages of their life cycle. We can see old, young and middle-aged galaxies, and work out the life cycle of an individual from these studies of the population at large.

All of this is a far cry from our knowledge of life on Earth. It was only in the 1980s that the idea that our entire planet could be regarded as a single living system, Gaia, began to be taken at all seriously, and the Gaia hypothesis is itself still the subject of intense debate. We have not yet really grasped the notion of a single planet as a living system, so it is small wonder that it needs a large leap of the imagination to regard the entire Milky Way Galaxy as a living system. And yet, the concept of Gaia, and Jim Lovelock's insight into the nature of life itself, provide just the springboard needed to make that leap.

Lovelock realized that the key feature of life on Earth is that the entire ecosystem of the planet is far from chemical equilibrium, and *stays* far from equilibrium. Even from a distance, a visitor from outer space could analyse the composition of the Earth's atmosphere, using spectroscopy, and infer that life was at work, creating *and maintaining* these non-equilibrium conditions. But a galaxy like the Milky Way is also in a far from equilibrium state, and is being maintained in that state by processes going on within it! The very criterion that led Lovelock to the insight that the entire Earth could be regarded as a living system applies to the Milky Way itself.

Of course, there are differences. The non-equilibrium state of the Milky Way is one that involves physical, not chemical, instabilities. And the timescale problem makes it hard for us to appreciate just how unstable the structure of the Milky Way ought to be. Nevertheless, it is far from equilibrium, and it is staying far from equilibrium, not settling down into what ought to be its natural, stable state.

The main clue is the spiral structure of the Galaxy. Like other spiral galaxies, ours is a flattened, disk-shaped collection of stars.

The disk is slowly rotating, with each individual star in the disk following its own orbit around the centre of the Galaxy. Because of the way gravity works, stars closer to the centre of the disk orbit more rapidly than stars farther out. This is like the way Mercury, the closest planet to the Sun, has a shorter 'year' than the Earth, while the giant planets, farther from the Sun than we are, have correspondingly longer 'years'.

In this situation, there should be no structure – no pattern – in the arrangement of stars within the disk. Any pattern that happened to form would quickly (within a few rotations of the Galaxy) get smeared out by the differential rotation. And yet, disk galaxies show clear and beautiful spiral structure, with arms of bright stars twisting out from the centre of the galaxy and bending gracefully back around it. The pattern is very similar to the swirling pattern of cream stirred into a cup of black coffee. But the pattern of white cream in black coffee in your cup is soon blended into a uniform brown colour by the differential rotation of the liquid. It is an unstable pattern, far from equilibrium. The same is true of the spiral pattern of stars in a disk galaxy like our Milky Way. Equilibrium would be a smeared-out distribution of stars in the disk, with no overall structure. And yet, studies of many galaxies show that the spiral patterns persist for many rotations of a galaxy's life.

If you had a lifespan of a few billion years, so that you could sit and watch galaxies rotate, and if you had a vantage point in the far depths of space, high above the Milky Way, looking down on its spiral pattern, you would see individual stars moving through the spiral pattern, following their orbits in obedience to the simple laws of physics. But the pattern itself, flying in the face of those simple laws, would not be disrupted as a result. Instead it would persist, refusing to be destroyed by the movement of the stars. It would be as obvious to you that something odd was going on as it would be to an alien spectroscopist that there is something odd about the atmosphere of the Earth. And as you watched the spiral arms grow, and move around the Galaxy seemingly independently of the differential rotation of the stars, it would be obvious that you were watching a living system.

The life processes that create and maintain the spiral structure in disk galaxies start with stars. A star like our Sun is itself, of course, in a state far from equilibrium. But even the keenest enthusiast for the Gaia hypothesis would not argue that the Sun is alive in the way that the Earth and the Milky Way are alive, because the Sun is doing the best it can to reach equilibrium. Inside the Sun, hydrogen is being converted into helium, and energy is being released. When all of the Sun's nuclear fuel has been exhausted, turned into heavier elements which represent a lower energy state, gravity will pull it together into a solid lump of cooling material. It will reach equilibrium at last, as a dwarf star in which the mass of the entire Sun (less all those 'elephants' turned into energy) is squeezed into a volume roughly the same as that of the Earth. But although an individual star cannot be regarded as alive, the processes that go on inside stars, and the way in which stars are born and die, are vital components in the processes which maintain the spiral structure of the Galaxy.

When a star like the Sun has burned all the hydrogen in its core to make helium-4, it has to adjust its structure to accommodate a new phase of nuclear burning. The core, now largely made of helium, shrinks slightly inwards, which releases gravitational energy and allows hydrogen burning to restart in a shell around the core. In the core itself, helium-4 nuclei – alpha particles – may combine in the triple alpha process, thanks to that all-important carbon resonance, to make carbon-12. Because the core is hotter, the outer layers of the star expand, and the star becomes what is known as a red giant. The Sun itself will become a red giant, swelling until its diameter is roughly the size of the Earth's orbit, about 5 billion years from now.

After this stage of its life, the fate of a star depends on its mass. Because of the 'failed' resonance at oxygen-16, many stars never get hot enough for much of their carbon to be converted into oxygen, and end up as dwarf stars made chiefly of carbon or helium, or both. But stars with rather more mass than the Sun get squeezed harder in the middle. They can produce a strong enough gravitational pressure in their hearts to drive more complex nuclear reactions. For example, the carbon-12 nuclei may collide and stick together to make magnesium-24, with energy being released. In violent collisions of this kind an alpha particle is sometimes ejected, so the end-product is a nucleus of neon-20. And sometimes two alpha particles are ejected, producing a nucleus of oxygen-16 and two alpha particles which can then take part in other fusion reactions. Neon-20 itself might gain energy from a gamma-ray and eject an alpha particle, leaving behind another oxygen-16 nucleus.

The whole complex web of interactions is called 'carbon burning', and takes place after all the helium in the core of a star has been used up. At a later stage neon-20 itself might 'burn', absorbing a stray alpha particle to make magnesium-24.

After carbon burning, a massive star will also begin fusion processes involving oxygen-16 nuclei, creating silicon, sulphur, phosphorus and magnesium; then finally silicon burning, in which nuclei of silicon-28 combine to form nuclei of iron-56. At every stage side-reactions convert some of these elements into those which are not made up of an exact number of alpha particles.

Iron-56 is the end of the line, though, because, of all possible arrangement of nucleons (protons and neutrons), the iron-56 nucleus has the lowest energy. Think of a mountain valley, with narrow, terraced fields up either side. On one side of the valley, the terraces represent the energy per nucleon stored in nuclei of hydrogen, helium and so on, going down the slope. On the other side of the valley, the terraces represent the equivalent energies of nuclei such as those of uranium and lead. At the very bottom of the valley lies the iron-56 level. Although all the nuclei up the sides of the valley can sit on their own terraces quite happily, if they are given a suitable nudge (even quite a small one) they will roll over the edge of the terrace and fall to the next level. Given enough nudges, they will fall all the way down to the iron-56 level. This is what happens as hydrogen is converted, step by step, into iron inside massive stars.

However, in order to make elements heavier than iron, the nuclei have to be given enough energy to climb up the other side of the valley, jumping from terrace to terrace. They have to be forced to fuse together, and they absorb energy in the process. In a sense, every nucleus would 'like' to become iron. Lighter nuclei are happy to stick together to make heavier nuclei, given the chance, because it moves them a step nearer to iron. Heavy nuclei, like those of uranium, can be relatively easily persuaded to split apart, forming fragments which are closer to the magic iron-56. And those heavier elements are made in the death-throes of stars much more massive than the Sun, when their cores collapse and gravitational energy forces nuclei together, in spite of the reluctance to move up from the bottom of the iron-56 valley.

Our Sun has spent about 4.5 billion years burning hydrogen into helium, and has about the same time left before it will become a red giant. But the more massive a star is, the faster it has to burn its nuclear fuel in order to provide the heat energy that holds it up against the inward tug of gravity. And each step in the nuclear fusion process provides less energy than the one before.

All of this is known from a combination of the study of many real stars at different stages in their life cycle, and calculations of how nuclear reactions will take place under the conditions inside stars, taking on board information from laboratory studies, such as the measurement of the energy levels in the carbon nucleus. To take just one example, this means that for a star about twenty-five times the mass of our Sun, typical of the kinds of star that form the bright spiral arms in a galaxy like our Milky Way, hydrogen burning lasts for only 7 million years, helium burning for a mere 500,000 years, carbon burning keeps the star hot for just six centuries, neon burning does the trick for about a year, oxygen burning is over in six months, and the conversion of silicon into iron-56 itself takes just one day. At that point, all hell breaks loose.

In fact, all hell breaks loose at this point in the life of any star with a mass more than eight times that of our Sun, although the inevitable is delayed longer for stars with masses at the lower end of the range. With a core made of iron-56, the most stable nucleus, the star has no way to generate heat to hold itself up against the pull of gravity. Quite suddenly, the core feels the full weight of all the layers of star stuff above it. The pressure is so great that electrons and protons are forced to blend into one another, merging to make neutrons. Although iron-56 is the most efficient way to pack a collection of neutrons and protons together, if you are dealing with neutrons alone the packing can be *much* tighter (partly because there is no longer any problem with the positive charges of the protons repelling one another). Within about a second, the entire core of the star, which is by now essentially a ball of iron nuclei as big as the Earth and weighing about one and a half times as much as our Sun, shrinks to a ball of neutrons – a neutron star – about the size of Mount Everest.

With the floor pulled from underneath it, the star stuff in the outer layers plummets down onto the new-born neutron star, accelerated by gravity to speeds as great as 15 per cent of the speed of light. When it hits the neutrons, all that energy of motion is bounced back the way it came, producing a hugely energetic shock wave that literally blows the star apart. For a star as big as 25 solar masses, at this stage of its life it will be a giant with a diameter as big as the orbit of Jupiter. Nearly 24 of the 25 solar masses of material will be above the neutron core. But the blast from the core is so powerful that it can shift all of this material. In the process, some of the energy of the blast creates those heavier elements that I have mentioned, and sends more than 20 solar masses of material, including carbon, oxygen, nitrogen and all the rest, out into space to mingle with clouds of hydrogen and helium and provide the raw material for the formation of new stars and solar systems.

It turns out that this is possible only because of another remarkable coincidence – another aspect of the Goldilocks effect. A critical ingredient in carrying the energy from the core into the shock wave which blasts a supernova apart comes from particles known as neutrinos. These are produced in profusion in the core under the violent impact of many solar masses of material falling onto it at a sizeable fraction of the speed of light. The readiness with which neutrinos interact with the star stuff is directly related to the strength of the weak force. If the weak force were a little weaker than it actually is, the neutrinos would pass right through the star stuff and escape into space without contributing any energy to the shock wave trying to blow the star apart. If the weak force were a little stronger than it is, the neutrinos would never get out as far as the shock wave, but would get involved in reactions down near the core itself.

Either way, the star would not be blown apart and there would be no spreading cloud of material to assist the formation of new stars. The weak force is actually 'just right' to ensure that supernova explosions make the maximum impact on their surroundings.

At the core of such an explosion, a relatively small supernova (with a mass of around 10 solar masses to start with) may leave the neutron star intact, to settle down into equilibrium as best it can. Many such neutron stars have been found and identified by the radio noise they emit; they are known as pulsars. But the more massive and violent supernova explosions may squeeze the inner core so much that it is crushed out of existence altogether, and becomes a black hole.

This happens to any neutron star which tries to form with a mass more than about three times the mass of the Sun. It also happens to any smaller neutron star which picks up enough matter from its surroundings to exceed this critical mass. Gravity is so strong for such an object that it bends spacetime around itself completely, shutting itself off from the Universe outside. The 'escape velocity' from inside this closed region of spacetime exceeds the speed of light, and therefore nothing at all can escape from it. It is described by Einstein's equations of the general theory of relativity, and in all respects it has become a universe in miniature, a world of its own. Unlike our Universe, however, which is expanding away from a singularity, the material inside a black hole made in this way collapses rapidly down into a singularity.

As we shall see in the next chapter, even that is not the end of the story, as far as the world inside the black hole is concerned; but it certainly is as far as our Universe, and the Milky Way Galaxy, is concerned. Once the black hole forms, we have no more direct contact with what goes on inside it. The birth of a black hole or a neutron star in a supernova explosion does, though, have profound implications for the life of our Galaxy. It not only provides the raw materials for life as we know it, including the carbon, hydrogen, oxygen and nitrogen, but it is also the basic life process that maintains the spiral structure of the Galaxy.

Once again, it is important to get this in the right perspective, in terms of time. By the standards of a human lifetime, supernovae are pretty rare events. The last supernova in our Galaxy that was seen from Earth was way back in 1604. The famous supernova of 1987 which caused a flurry of excitement among astronomers was the nearest one seen since then, and actually occurred in a small galaxy that is a near neighbour of the Milky Way. Of course, we cannot see every star in the Milky Way, and some supernovae must explode in our Galaxy without being noticed by terrestrial astronomers. From studies of other galaxies, astronomers estimate that a supernova goes off somewhere in our Galaxy every few tens of years. There are roughly two every century.

That still does not sound like much, in a galaxy containing roughly a hundred billion stars. But the lifetime of the Galaxy is so vast that by comparison this is a very rapid process. If there are two supernovae every century, then there will be twenty in every millennium, and 20,000 in every million years. What is more, supernovae were much more frequent when the Galaxy was younger and full of gas and dust that had not yet been converted into stars. Since the Sun was born, 4.5 billion years ago, our Galaxy has experienced several hundred million supernova explosions. And the gas at any point in the disk of our Galaxy is exposed to the blast from a supernova at least once every few million years. On the timescale of galaxies, this is a frequent event. Although we cannot see it, the structure of the gas that lies between the stars must be a kind of foam of coalescing supernova remnants. But at any instant in the life of a galaxy like our own, most supernova explosions are taking place along the trailing edges of the spiral arms themselves.

Indeed, the spiral arms show up so prominently in photographs of spiral galaxies only because they are edged by hot, young, blue-white stars. These stars, much more massive than our Sun, have lifetimes of only a few (up to ten) million years, but shine very brightly during their brief lives. Most of them end their lives in supernova explosions. The density of material (stars, gas and dust) in a spiral arm is only slightly greater than in the rest of the disk of the galaxy; what is different is not so much the density of matter available, as what the galaxy is doing with it.

Unlike the spiral pattern of cream stirred in your coffee, these arms are not simply random features being 'wound up' into a spiral pattern by differential rotation. In round numbers, a galaxy rotates once every hundred million years or so. We know this from yet another application of spectroscopy, using the Doppler effect to measure shifts in the spectral lines of stars in spiral galaxies and work out the speed at which those stars are orbiting the centre of their parent galaxy. But the lifetime of a galaxy is measured in billions of years, so that any typical spiral galaxy has had time to rotate scores of times since it was born. If the spiral arms were simply a result of the differential rotation, they would be wrapped around the galaxy in tight spirals, with about as many turns as the number of times the galaxy has rotated. Yet, in fact, the typical spiral pattern shows just one or two turns in the arms, traced by a line of bright stars all the way from the centre of the galaxy to the edge of the disk.

The explanation is that the spiral arms are the visible part of a density wave, which is moving round the galaxy in the same direction as the stars themselves, but more slowly. This is rather like a wave on the surface of water. In the water, individual molecules bob up and down (they actually move in small circles) as the wave passes, but they do not move forward with the wave.

Although the wave looks like water moving physically across the surface of the ocean, it is made up of different molecules at different times.

You can understand the spiral density wave of a galaxy by imagining looking down on the galaxy from above. Each star orbits around the centre of the galaxy in an almost perfectly circular orbit. But it also weaves in and out a little in its orbit, towards and away from the galactic centre, because of the gravitational influence of other stars and dark matter in the galaxy. This weaving in and out means, in effect, that the star is moving in a small circle, while the *centre* of that circle follows the circular orbit around the galaxy, like a moon orbiting a planet going round the Sun. This makes the overall motion of hundreds of millions of stars combine to form a slow-moving spiral density wave.

There is a nice analogy which helps to explain the appearance of such a moving density wave. Imagine a city, like London, completely encircled by a motorway (the M25). If the motorway were filled with cars moving in the same direction at a steady speed of, say, 100 kph (60 mph), then at night-time someone flying overhead in a helicopter would see a pattern of lights circling smoothly around the motorway. But now suppose that there is one slow-moving vehicle on the motorway, a large truck moving in the same direction as the other traffic, but at only 50 kph. As the other vehicles come up behind the truck, they will have to slow down while they find a slot in the traffic in the outside lane to enable them to overtake. There will be a bunch of cars behind the truck. The actual cars in the bunch will be constantly changing – the ones at the front pulling out to overtake and proceed on their way at 100 kph, while new cars join the queue at the back. But what you will see from the helicopter flying over the motorway is a stream of lights moving steadily at 100 kph, with a dense knot of lights (a density wave) moving along more slowly, at 50 kph. And a photograph will, of course, freeze the motion, showing simply a circle of evenly spaced lights, with one knot of densely packed lights just behind the position corresponding to the truck.

Because spiral galaxies rotate so slowly, on the human timescale, we effectively see them frozen in time in this way, even though exactly the same sort of dynamic activity is happening in the spiral arms of a galaxy like our own. The density wave in a spiral galaxy moves around the galaxy at a speed of about 30 km per second, about three times faster than the speed of sound in the tenuous gas of interstellar space. This means that it produces a supersonic shock wave, like the shock wave associated with a supersonic aircraft such as Concorde, at the leading edge of the spiral arm (the outer, convex part of the spiral). But the stars and clouds of material in the galaxy are moving through the shock wave with speeds of 200–300 km per second. Clouds of gas and dust pile up in a traffic jam just behind a spiral arm, along the inside bend of its curve. They are squeezed into the shock wave, and this squeezing leads to the birth of giant stars.

Once something happens to set this process off, it is very largely self-sustaining. The explosion of a supernova, sending a blast out through the surrounding interstellar medium, is just the thing to squeeze any nearby clouds that are on the brink of collapsing to form new stars, setting off a further burst of star formation. Once the process gets going it can spread, like a forest fire, across the affected region of a galaxy. And remember that, like those cars overtaking the truck on the motorway, *all* the gas clouds in the galaxy pass through the spiral arms – one on each side of the galaxy, so twice in every orbit. Every hundred million years or so, everything in the galaxy is subject to this squeeze. As the spiral waves cycle around the galaxy, they recycle interstellar material. Old gas and dust left behind from old, dead stars gets squeezed into new stars.

The clouds that collapse to form giant stars along a spiral arm today are, though, very different from the simple mixture of hydrogen and helium gas out of which the first stars were born, not long after the Big Bang. Although the clouds are still mostly composed of hydrogen and helium, they also contain other molecules, including water and organic (carbon-based) compounds such as formic acid, formaldehyde and ethanol, plus a trace of dust, in the form of carbon grains. The Orion Nebula, a site of star birth relatively near to us, in the constellation Orion, contains at least sixty different kinds of molecule, as identified by their spectra. This is typical, although the Orion Nebula happens to have been studied in detail because of its proximity to our Solar System.

Altogether, astronomers estimate that the total amount of molecular gas in our Galaxy inside the orbit of the Sun may be as much as 3 billion solar masses, 15 per cent of the total mass of stars in the same region. Cloud stuff is constantly being turned into stars, and star stuff is constantly being recycled to produce new clouds.

A typical molecular cloud has a diameter of about a hundred light years, with a mass up to several million times that of the Sun. When it does collapse, it produces not just one or two giant stars, but a profusion of them. Each cloud spawns a bright blue clump of stars, and these star clusters are dotted along the spiral arms, so that when we see a spiral galaxy from a distance the effect is like sapphires strung out on a spiral necklace.

It is harder than you might think, though, for such a cloud to collapse and fragment to form stars. The problem is that when a cloud shrinks, gravitational energy is released, and this heats up the molecules in the cloud, making them move faster so that they collide with one another more vigorously, producing a pressure which stops the shrinking. The cloud has to find some way of getting rid of this energy, in the form of radiation, so that it can cool down and collapse further, down to the densities at which stars are born. This happens in two ways. First, many of the molecules in the cloud, such as carbon monoxide or water vapour, radiate energy when they get hot. We know this process is going on because it is the characteristic radiation produced by these hot molecules that enables us to detect them. Secondly, when stars begin to form in the densest region of a giant molecular cloud they will produce a lot of visible light and ultraviolet radiation. This would tend to blow the cloud apart and stop the processes of star formation, were it not for the presence of grains of carbon dust in the cloud. Even though the dust may amount to no more than 1 per cent of the cloud's mass, it absorbs the ultraviolet radiation and re-radiates it in the infrared part of the spectrum, at wavelengths at which the energy can escape more easily into space. Many observations show that regions of strong infrared emission are also regions of active star formation. Supernova remnants, in the form of expanding shells of gas, are also associated with regions of active star formation.

The details of all these processes are still only poorly understood, and the investigation of molecular clouds, spiral structure and the formation of short-lived blue-white giant stars that explode as supernovae is one of the most important and exciting areas of astronomical research today. But it is already clear that carbon dust grains, as well as complex molecules containing carbon, play a vital role in the cooling process without which the stars that edge the spiral arms would not form in such abundance. It is even possible that the kind of complex organic chemistry (that Fred Hoyle, for one, would identify with life processes) going on in interstellar clouds far from any planet plays a role in the processes of star formation. The processes by which stars form, and galaxies evolve, are more complicated than might have been expected, and Lee Smolin, of Syracuse University in New York State, has gone so far as to say that:

> there seems to be a kind of ecology in the physics of spiral galaxies by means of which the structures responsible for star formation – the spiral arms and the associated clouds of dust and gas – are maintained for time scales much longer than the relevant dynamical time scales . . . These must involve self-organizing cycles of materials and energy of the kind that one sees in diverse non-equilibrium states as well as in biological systems.

The point is that the influence of supernova explosions, spreading both heavy elements and shock waves through the interstellar medium, is essential for the creation of conditions which allow the kind of stars that will end their lives as supernovae to form. This is nothing less than Gaia on a galactic scale. And it all depends on the delicate balance of the forces of nature – the Goldilocks effect. The weak force has to be 'just right' to allow supernovae to spread their material out into space. Carbon, which exists only because of the coincidences discussed in Chapter Seven, is necessary to the process, and complex organic molecules are probably necessary as well.

All this puts those coincidences into a very different perspective. The Universe is not, after all, set up for the benefit of organic life-forms like ourselves. It is set up so that galaxies like our Milky Way can operate as supernova nurseries, producing successive generations of massive, short-lived stars in association with spiral arms. It just happens that these processes require carbon, and the presence of complex organic molecules, and water. So the production of carbon, organic molecules and water has evolved as part of the life processes of galaxies. Once that carbon, those organic molecules and water exist, it was probably inevitable that life-forms like us should evolve. But we are best seen as a by-product of the processes by which our Galaxy maintains itself in a far from equilibrium state – the processes by which it keeps itself alive.

There is another oddity about the spiral pattern. Although the natural form of disturbance that will grow in a disk galaxy is indeed a spiral density wave, many calculations and computer simulations show that the resulting spiral pattern should last only a few rotations of the galaxy. So why do so many of the galaxies we can see today show spiral structure? The computer modellers discovered that the simplest way to make the spiral pattern persist for much longer is to include the gravitational influence of an invisible halo of dark matter surrounding the galaxy. If the mass of the halo is ten times the mass of all the bright stars put together, it stabilizes the spiral pattern and makes it last much longer, before eventually changing its shape to form a bar across the centre of the galaxy.

We do indeed see such barred spirals, as they are called, in about the right proportion to agree with this calculation. Astronomers have also inferred that spiral galaxies are embedded in dark haloes from measurements of the way in which the disks rotate – the outer regions of the disk rotate faster than they would if they were not embedded in such a halo. And, of course, both cosmological theory and the COBE observations tell us that the Universe contains up to a hundred times more matter than we see in the form of bright stars and galaxies. Not only does everything hang together rather well, but the dark matter (or at least some of it) seems also to be an essential ingredient in ensuring that spiral galaxies operate as efficient supernova breeding grounds.

The processes of a self-regulating system of large-scale star formation have been discussed by Guillermo Tenorio-Tagle, of the Max Planck Institute for Astrophysics, in the authoritative *Astronomy and Astrophysics Encyclopedia* (1992). The striking feature of the way in which spiral galaxies maintain a steady state, far from equilibrium, is that massive stars which live for only one to ten million years each, and are formed from clouds which collapse to form new stars on timescales of a hundred thousand to a million years, are able to maintain a spiral pattern which persists for a billion years or more. In the process, the rate at which dust and gas is converted into new stars is almost the same as the rate at which old stars return matter to the interstellar medium – in both cases, a couple of solar masses of material per year throughout the Galaxy. It does not sound like much, but it adds up dramatically over a few million years.

The whole process is reminiscent of the way in which a collection of cells (a thousand times more cells in your body than there are bright stars in the Milky Way Galaxy), many of which have a lifetime of only a few weeks, are able to maintain the pattern of your body for seventy years or more. It is estimated that your body 'loses' over a kilogram of cells every day, chiefly from your skin and the lining of your gut – but you do not 'wear out', because these cells are constantly being replaced. Neither does the spiral structure of a galaxy like the Milky Way 'wear out', because the bright, blue stars are constantly being replaced.

Tenorio-Tagle describes the processes by which the interacting shock waves from supernova explosions create new giant molecular clouds as one of 'continuous reproduction', pointing out that the process is indeed 'self-regulating'. In a cloud where a few supernova explosions occur, the spreading shock waves from the explosions will gather up interstellar material to make a larger cloud, which will collapse to make even more giant stars. This, presumably, is how the whole process got started, when our Galaxy was young. But the process of molecular cloud growth cannot go on forever. If it reaches the point at which star formation proceeds too fast in a collapsing cloud, because it is denser than the optimum, the energy radiated by the first young stars will blow the cloud apart and stop the processes of star formation early. So the next generation of clouds will be less dense, and therefore more efficient at making supernovae at the right rate. The clouds in which conditions are just right for the formation of many stars will produce many supernovae, leading to many overlapping shock waves that in turn produce many new clouds that are just right to be the nurseries in which the next generation of stars can form. And if a cloud is either a little too big or a little too small, the natural feedback processes will make its descendants, in the next generation of clouds, more nearly 'just right'.

Crucially, it is the explosion of many massive stars close to one another that forms large new clouds (several hundred light years across) as the shock waves sweep large amounts of matter together. So clouds which are 'just right' for the birth of many supernovae produce supernovae which are 'just right' for the production of clouds similar to the ones that gave them birth. 'The selective evolution will proceed', says Tenorio-Tagle, 'in an almost "Darwinistic" manner.'

But why quibble? Why put in that qualifying 'almost'? This *is* evolution by natural selection and survival of the fittest, in the strict Darwinian sense, and the fact that it involves stars and clouds of gas and dust hundreds of light years across, and timescales of millions of years, all going on in a spiral galaxy, should not blind us to the truth. There is surely no need for that 'almost': the processes of star formation in spiral galaxies today are so efficient because they have indeed evolved to make best use of the available materials. It may be stretching a point to argue that an individual star is alive; but it is also stretching a point to argue that our Galaxy is *not* alive, in the same sense that the Earth is alive. And within the living Galaxy a typical giant molecular cloud/supernova system seems every bit as alive as a typical caterpillar/butterfly system here on Earth.

Not that all galaxies are equally efficient at breeding supernovae. As you might expect, there are some in which there is no obvious spiral pattern, and the 'forest fire' effect seems to have run wild, sweeping through the entire galaxy, producing bursts of star formation everywhere. At the other extreme there are the so-called elliptical galaxies, lens-shaped systems in which very little star formation seems to be going on today. But the modern understanding of the way in which these galaxies have come to be as they are also owes more to ideas that seem to come from the realm of biology than from good old-fashioned physics. The description of the evolution of galaxies is littered with words (like 'evolution' itself) borrowed from biology, and although the astronomers hasten to reassure us that the terminology is intended as nothing more than metaphorical, it increasingly seems that they are not fooling anybody except themselves.

Astronomers' ideas about the origins and evolution of galaxies changed dramatically in the early 1990s, as improved telescopes (including the Hubble Space Telescope) and other instruments enabled them to probe further back in time by studying fainter, more highly redshifted galaxies. They used to think that the galaxies we see around us today all formed at more or less the same time, just after the Big Bang, and have since changed internally as they have aged, while together presenting much the same overall picture of the Universe at large. This notion has now been discarded completely, to be replaced by a dynamic, evolving picture in which galaxies compete with one another, merge, and absorb one another. The most striking change is that the elliptical galaxies, once thought to be the oldest, are now seen as relative newcomers.

Elliptical galaxies are less numerous than spiral galaxies, but the biggest of them are much bigger than even the biggest spiral. While spirals weigh in at around a hundred billion solar masses, the largest ellipticals may be a hundred times bigger. At the other extreme there are swarms of dwarf galaxies, whose numbers are difficult to estimate because they are faint and hard to see. A dwarf galaxy may be no bigger than a cluster of stars within the halo of the Milky Way Galaxy, less than a kiloparsec across and with a mass of around a million Suns.

Elliptical galaxies are made up mainly of cool, red stars. Since cool, red stars are old, astronomers thought this meant that the galaxies had formed long ago. Elliptical galaxies also contain little gas and dust. But spiral galaxies do also contain old, red stars, not just the young stars typical of the disk. The red stars are found both in the central bulge of such a galaxy (the 'yolk' of the 'fried egg') and scattered around the visible halo. It is now clear that the red stars in elliptical galaxies have actually come *from* spiral galaxies. Ellipticals form from the merger of two spiral galaxies that collide with each other, and from events in which a spiral galaxy is absorbed by an elliptical galaxy. This is how ellipticals get to be so big.

The evidence comes both from observations of galaxies in the act of colliding and from computer simulations of collisions. When two spiral galaxies collide, the computer models show, the thin disks are destroyed and become merged into a single star system with a shape just like that of an elliptical galaxy. In such a collision between galaxies the stars themselves do not collide with one another – remember how the chocolate-sweet model demonstrated the vast distances between stars. But as two galaxies merge, the gravitational fields interact to pull all the stars into one system. The clouds of gas and dust in the two galaxies really do collide, sending shock waves rippling through the new system.

When a larger elliptical galaxy swallows up a smaller spiral galaxy, the elliptical grows, but many stars from the spiral end up following similar orbits to one another inside the elliptical.

Photographs of ellipticals show bright arcs of stars within them, corresponding to these crowded orbits. Ellipticals, which used to be thought of as featureless star-piles, are also shown by modern instruments to be criss-crossed by faint ripples of light, stripes and crosses, all of which provide unambiguous evidence of spiral galaxies that have been swallowed up by the ellipticals and are still being digested.

Although it gets harder to see details of individual galaxies the farther out into space (the farther back in time) we look, whole clusters of galaxies can be monitored more easily at large redshifts. Large clusters near us contain many ellipticals and are distinctly reddish. At distances corresponding to a 'look-back time' of about five billion years, however, the clusters are much bluer, which shows that active star formation was going on in them at that time. The Hubble Space Telescope shows that the blue objects in many of these clusters are actually pairs of spiral galaxies in the process of merging with one another to make ellipticals. It seems that the gas and dust clouds in the spirals are triggered into a final burst of star formation – a last great forest fire – by the gravitational disturbance and the shock waves associated with the merger.

That is why elliptical galaxies today contain little or no gas and dust and very few blue stars: because all the gas and dust was turned into blue stars when the ellipticals were formed in mergers between spiral galaxies, a few billion years ago. Almost certainly, our Galaxy will eventually merge with the spiral galaxy M31 now seen in the constellation Andromeda, forming a new elliptical galaxy in a blaze of blue-white glory. But the speed with which a newly forming elliptical galaxy settles down into a stable pattern, with the stars from different galaxies ordered into a new arrangement, is further strong evidence that all galaxies are embedded in massive dark haloes. It is the gravity of the dark matter that pulls everything together so neatly and so quickly.

The largest elliptical galaxies of all are found at the centres of clusters of galaxies. There they sit, as some astronomers have put it, 'like a spider in its web', and become fat by consuming any smaller galaxy whose orbit through the cluster takes it too close to the supergiant. In the 1990s, astronomers describe the way in which elliptical galaxies form using words like 'devour', and sometimes refer to the elliptical that comes off best in a merger between two galaxies of unequal sizes as a 'cannibal' galaxy.

Although less than 1 per cent of the galaxies we can see today are actively involved in mergers, these processes are so short-lived, on a universal timescale, that it seems likely that half of all the galaxies in the Universe have merged with galaxies of similar size in the past seven or eight billion years. Mergers were certainly even more frequent during the first half of the life of the Universe to date.

Spiral galaxies themselves also now seem to have formed from the merger of smaller units – or rather from competition among smaller units, which have fed off each other and grown in the process. The globular clusters that surround our Galaxy are now regarded rather like fossils here on Earth. Their ages can be measured quite accurately, from analyses of their chemical composition, and it turns out that, for example, while one cluster is 14 billion years old another is just 10 billion years old, and a third a mere 7 billion years old. The spread of ages shows that our Galaxy formed from an amalgamation of about a million smaller gas clouds, each weighing up to about a million times the mass of our Sun.

Whenever a new gas cloud collided with the growing Galaxy, the shock wave would trigger a burst of star formation, producing a new globular cluster. The remaining gas from the cloud would then be tugged by gravity and slowed down by friction to add its contribution to the rotating disk. The first generation of hot blue stars would have quickly run through their life cycles and exploded at the beginning of this process, lacing the disk with heavy elements. Molecular clouds would have formed, and soon the characteristic spiral arms would have emerged.

This whole process of mergers between smaller units to build up galaxies like the Milky Way works particularly well in the context of the cold dark matter scenarios, where the gravity of the dark matter helps to draw the pieces together. Indeed, without dark matter it would be very difficult to explain how galaxies formed at all. Once again, the COBE observations have confirmed what astronomers knew had to be the right picture of the Universe.

Writing in *New Scientist* in 1992, Nigel Henbest describes a past scene:

> If we could view the Universe soon after the big bang, we would not see young versions of spiral and elliptical galaxies. We would, instead, be surrounded by umpteen small clouds of gas, jostling against each other. Rather as single cells evolved into the complexity of life on Earth, these small gas clouds have given rise to all the diversity of the galaxies in the Universe today, from the beautiful spirals to the supergiant ellipticals.

Again, that qualifying 'rather as'! Surely this whole process *is* an example of evolution at work, with competition among individuals for natural resources (gas with which to make stars) leading to the growth of more complex systems.

The whole picture is so new, however, that there are still puzzles to be explained and holes to be filled in to complete the story of the evolution of galaxies. One concerns the mystery of dim, blue galaxies first identified by Anthony Tyson and his colleagues at AT&T Bell Laboratories in the 1980s. There are huge numbers of these dwarf galaxies at redshifts corresponding to an epoch 2 to 3 billion years ago, when the Earth was about half its present age. Had there been astronomers alive on Earth then, their telescopes would have shown the Universe blazing with the light of these blue dwarf galaxies. Each one was just one hundredth the size of the Milky Way, and there were so many of them that their appearance on long-exposure astronomical photographs has been described as 'cosmic wallpaper'; but nothing like them is active today.

Where have they gone? It would be nice to think that they represent an intermediate stage in the formation of galaxies like our own, and have been eaten up by the spirals we see today; but it is hard to see how the thin disk of a spiral galaxy could survive mergers involving such relatively large objects. The dwarf galaxies may simply have burned out. Being so small, a firestorm of star formation and resulting supernovae could have produced a shock wave powerful enough to blast all their remaining gas and dust out into space, escaping entirely from the weak gravitational grip of the dwarf – in which case the dwarfs are still there, but now too faint to see.

Before the era of the blue dwarfs, the Universe was populated by galaxies that were larger in size than the galaxies of today, but not larger in mass. They were each more spread out, because gravity had not had time to pull the various pieces from which they were growing together.

In an article in the journal *Science*, commenting on these new developments in early 1992, the writer was a little less coy about what may be going on than Nigel Henbest:

> These days, an outsider listening to some astronomers discussing their subject might mistake them for biologists discussing an ecosystem. Their conversation is full of references to populations that are emerging, evolving, and even going extinct. But instead of living things, they are talking about galaxies.

Again, though, why drop in that 'instead of'? The article goes on to say that 'whole groups of galaxies have evolved together like living species'. I would say 'as' instead of 'like'. The reason why the language of biology is more appropriate than the language of physics to the description of the evolution of galaxies is that galaxies are indeed behaving as living systems. And in at least one respect the view of a galaxy like our Milky Way as a larger-scale version of Gaia has an advantage over the original idea of the Earth as a living system.

Critics of the Gaia hypothesis sometimes complain that as we know of only one living planet, and can see no signs of competition between planets for resources, the extension of evolutionary ideas to Gaia is at best imprecise and at worst wrong. Since we can see millions of galaxies, all exhibiting the Gaian behaviour of maintaining stability far from equilibrium *and* competing with one another so that only the fittest survive, that criticism cannot be applied to Gaia on the galactic scale.

A few astronomers, among them Sir Martin Rees of the University of Cambridge, are prepared to begin to talk about taking on board ideas from biology to apply to the study of the Universe at large. They are willing to use concepts like ecosystems and population dynamics, while being careful to stress that these are 'only metaphors', even if they do seem to work rather well. This caution is nothing new. Jim Lovelock himself sometimes points out that the idea of Gaia can be useful as a metaphor even if it is not strictly accurate in biological terms. He argues that the concept provides new insights into feedback processes at work here on Earth, such as the interactions between micro-organisms in the sea, dimethyl sulphide, cloud cover and ice ages. This seems something of a cop-out to me. But if it gives the next generation of astronomers enough reassurance to make use of these ideas, if only as metaphors, then eventually they may become so used to them that they will drop the qualifications.

It certainly is important that the idea of evolving populations should be taken seriously in an astronomical context, because on the largest possible scale they now seem to provide the only real insight into the origin and nature of the Universe itself. But before I can begin to address those questions, one feature of the Universe still needs to be explained.

I have argued that our Galaxy is alive – literally alive, in the full biological meaning of the term. Like other galaxies it has been produced by a process of evolution and competition within the Universe, following the Big Bang. The end-product of this evolutionary process has been spiral galaxies that are very efficient supernova nurseries. When spiral galaxies themselves merge, or are eaten, to make elliptical galaxies, there is a final blaze of supernova activity that processes virtually all of the remaining gas and dust in the system. The question is, why should galactic evolution favour the occurrence of supernovae?

The answer lies in what happens to the core of a supernova

when the star itself explodes. In very many cases it becomes a black hole. In other cases it becomes a neutron star – and some of those neutron stars themselves will swallow up enough extra matter to become black holes. As of 1992, astronomers have identified a handful of such 'stellar mass' black holes in our Galaxy. By their very nature, black holes, being black, are hard to spot. But a few that happen to be in binary systems, orbiting a more normal star, reveal their presence by tearing off gas from the companion star with their strong gravitational pull, and swallowing it. On its way down the throat of a black hole, the infalling matter gets heated to very high temperatures, and radiates X-rays. X-rays coming from an otherwise invisible companion orbiting a normal star are a strong indication of the presence of a black hole. Some of these 'binary X-ray sources' are neutron stars, but in other cases the dynamics of the orbit of the binary system show that the unseen star has a mass greater than three times that of the Sun, and so must be a black hole.

To get some idea of how many of these stellar mass black holes there might be in the Milky Way, astronomers make two independent calculations. One is based on the number of supernovae there have been during the lifetime of the Galaxy to date. The other extrapolates from the tiny sample of X-ray sources surveyed so far to the whole Galaxy, assuming that the same proportion of black holes found so far will be found in the Galaxy at large. Both estimates give the same figure. In round terms, there are several hundred million black holes in our Galaxy alone.

There are millions of other spiral galaxies in the Universe today, each carrying a similar cargo of black holes. And the extreme firestorms in which spiral galaxies merge to create ellipticals produce many more millions of black holes in the supernovae that consume the last traces of gas and dust in those galaxies. So a giant elliptical galaxy may well contain tens or even hundreds of billions of black holes. The Universe is extraordinarily good at turning matter into black holes, and the process depends, as we have seen, on a chain of quite extraordinary coincidences in the laws of physics. This realization is the key to the new understanding of the Universe itself – that it, too, is alive, and that it has evolved in competition with other universes.

CHAPTER NINE

The Living Universe

The central mystery in cosmology is the uniformity of the Universe: everywhere we look, it presents the same overall appearance. Of course, there are different galaxies, arranged in slightly different patterns, seen in different parts of the sky. But whichever part of the sky we choose to look at, or photograph, we see the same sorts of galaxy arranged in the same sorts of pattern. Overall, the Universe really is remarkably smooth and uniform. The most extreme example of this, and the clearest indication of why it is a mystery comes from looking at the cosmic microwave background radiation.

Remember that this weak hiss of radio noise, which we detect today at a temperature of just under 3 K, comes to us from a time when the Universe was just 300,000 years old. At that time, when the temperature everywhere in space had cooled to about the temperature of the surface of the Sun today (roughly 6000 K), electrons were able, for the first time, to link up permanently with nuclei of hydrogen and helium to form stable atoms. With all the electrically charged particles locked away in electrically neutral atoms, electromagnetic radiation (light and radio waves) could no longer interact with matter. The Universe suddenly, for the first time, became transparent to radiation; all the radiation could do for the next 15 billion years or so was cool gradually as the Universe expanded, all the way down from 6000 K to the 3 K that we can detect today.

What we see when we look at the background radiation is an image of what the Universe was like the last time that radiation interacted with matter, 300,000 years after the Big Bang. This is rather like the view we get of the Sun. What we see when we look at the Sun is an image of what the Sun is like at the point where radiation last interacted with matter, at the surface of the Sun. The difference is that the 'surface' the background radiation comes from is a surface in history, not a surface in space. All kinds of interesting things go on inside the Sun, but the light that reaches us carries the imprint of conditions on the surface, not those in the deep interior. In the same way, all kinds of interesting things happened in the first 300,000 years after the Big Bang, but the background radiation carries the imprint of conditions at that time, not of the Big Bang itself.

In both cases, though, what we see does tell us something about those deeper, or earlier, conditions. Apart from minor fluctuations, the Sun radiates the same amount of energy out into space in all directions, which tells us that it is a fairly uniform object inside. The background radiation is *very* much the same in all directions, which tells us that the Big Bang itself was very smooth and uniform.

This is the single most important fact about this radiation (apart from the fact that it exists, providing direct evidence that the Universe was born in a hot Big Bang). It is uniform. When COBE first looked at the background radiation, it almost immediately pinned down its temperature to 2.735 K, the most accurate measurement of it made up to that point. It found that the spectrum of the radiation *precisely* matches the pattern of a so-called black body spectrum (the spectrum predicted by the Big Bang model), to an accuracy of better than 1 per cent. And it showed that the differences in temperature of the radiation coming from different parts of the sky is less than one part in 10,000. In other words, the Universe was smooth to more than one part in 10,000 at a time 300,000 years after the Big Bang.

The big puzzle about all this is how the radiation from one side of the sky 'knows' the temperature of the radiation from the other side of the sky, so that it can match it. Regions on opposite sides of the sky last interacted (last 'touched' each other) way back in the Big Bang itself. Since then they have been carried farther and farther apart by the expansion of the Universe – and that expansion acts to make any irregularities bigger as time passes, not smaller. What is worse, it turns out that regions on opposite sides of the sky today *never were* in contact with one another, even in the Big Bang (at least, not according to the simplest version of the Big Bang model).

If you wind back, in your imagination, the expansion of the Universe to a tiny split second after the moment of creation, everything we can see would have been crowded into a volume of space just one millimetre across. It sounds ridiculous, but that is what the observations of the expanding Universe and the equations of the general theory of relativity tell us. That tiny seed would have contained all of the energy that later went into making up stars and galaxies and dark matter (in line with $E = mc^2$, or rather, in line with $m = E/c^2$), and the radiation that would become the cosmic background. The time when the Universe was just under a millimetre across corresponds to a moment 10^{-35} of a second after the moment of creation itself. Over the ensuing 15 billion years, that millimetre-sized 'seed' grew to become the entire visible Universe, with a diameter some 10^{28} times greater.

But this poses an enormous puzzle. Remember that light travels at a speed of 3×10^{10} cm per second – nothing can travel faster, so this is the fastest that any message can travel through space. When the Universe was a mere 10^{-35} of a second old, then, no signal of any kind – no information – could have travelled more than a distance of 10^{-25} cm. (In round terms, we can ignore the '3' in the speed of light.) A volume of space just one millimetre (0.1 cm) across at that time would have been made up of 10^{24} separate regions, none of which could have had any knowledge at all about what was going on in the other regions. There literally had not been enough time for one of these domains, on one side of the millimetre-sized 'seed', to find out what was going on in another domain on the other side of the seed. And as the Universe expanded, each of those domains would have grown to about 10 metres across today. Nearby domains would have merged and shared information, in the form of radiation travelling at the speed of light, but they should have produced a distinctly grainy picture, like a greatly over-enlarged photograph, at the time when the Universe was 300,000 years old. Yet there is no trace of this graininess in the background radiation.

The puzzle was resolved in the early 1980s, by Alan Guth, of MIT. I chose the time 10^{-35} of a second after the moment of creation in this example of the astonishing smoothness of the Universe because in Guth's resolution of the puzzle there was a dramatic change in the nature of the expanding Universe around that time. The resolution is the theory that goes by the name of inflation, which I touched on in Chapter Two.

Inflation describes the processes that gave rise to the distinctions between the fundamental forces of physics. The best theories of physics that we have are the so-called 'grand unified' theories. They are the best because they explain the most observed details of how the forces of nature work; even under the extreme conditions produced when beams of particles are smashed into one another inside particle colliders like those at CERN, at Geneva. In many ways, such collisions resemble, for a fleeting instant, the conditions that existed in the Big Bang itself. And if a theory works as an accurate description of conditions in particle experiments which *nearly* resemble the early phases of the Big Bang, the same theory can be used with numbers corresponding to still higher energies to calculate what went on even earlier in the Big Bang itself. The process is known as extrapolation, and it is not perfect. There is always a chance of some nasty (or pleasant!) surprise turning up at higher energies. But so far it has passed every test with flying colours – and the latest 'atom smashing' experiments are getting to energies which only existed naturally during the first second of the existence of the Universe.

It is worth taking a sneak preview of the kind of experiments planned for the late 1990s. If the Universe was born out of a hot fireball of energy, as the Big Bang theory tells us, how did that energy get converted into the matter that we see around us today? The standard theory of matter says that ordinary protons and neutrons – members of a family of particles sometimes called hadrons – are composed of fundamental entities known as quarks, held together by swapping whimsically named 'gluons' between themselves. The exchange of gluons produces a force so strong that no individual quark can ever escape from a hadron. But under the conditions of extreme pressure and temperature during the first split second of the Universe, 15 billion years ago, individual hadrons could not have existed. Instead, according to standard theory, the Universe consisted of a soup of quarks and gluons – a 'quark–gluon plasma'.

The quark–gluon 'era' ended about one-hundred-thousandth of a second after the Universe began expanding from a point. At this critical time, a phase transition took place – equivalent to the way steam changes into liquid water – and hadrons were formed. This was an 'everyday' phase transition, very similar to the ones that go on when steam condenses to make water today, and had nothing to do with the inflationary process. But this phase transition, unlike steam condensing into water, happened everywhere in the Universe at once. Physicists on both sides of the Atlantic are now planning experiments which will probe the quark–hadron transition, providing further experimental tests of the theories on which our understanding of the early Universe are based.

To get a feel for just how extreme the conditions are, we need to look at temperature and density in terms rather different from those of everyday life. Physicists measure both quantities in the same unit – the electronvolt, or eV. Strictly speaking, this is a measure of energy, so it is a perfectly good measure of temperature. Particles which collide with one another with kinetic energies of a few electronvolts have a temperature equivalent to a few tens of thousand degrees, on the Kelvin scale, and these are the energies and temperatures associated with ordinary chemical reactions.

Energy can be converted into a mass equivalent by dividing by c^2, in line with $E = mc^2$, and when electronvolts are used as units of mass, or in expressing densities, the division by c^2 is taken as read. In these terms, the mass of an electron is 500 keV, and the mass of a proton is 1 GeV. The neutron has almost the same mass (actually slightly more, which, as we saw in Chapter Eight, is one of the key Goldilocks effect coincidences), and the packing of neutrons and protons together in atomic nuclei, or neutrons in a neutron star, provides the greatest density of matter that can exist in the Universe today (except for a faint possibility that hadrons may be squeezed so hard in the centres of some neutron stars that they are squashed into a quark–gluon soup). The radius of a proton is about 8×10^{-16} m), which is near enough, for our present needs, to one femtometre (1 fm = 10^{-15} m). So the density of a proton – the ultimate density of everyday matter – is, in round terms, 1 GeV per fm^3.

Calculations of how matter behaves at the quark–hadron transition have been carried out using powerful computers. The critical temperature for the phase transition (corresponding to the critical temperature at which water boils) is in the range from 150 to 200 MeV, according to these calculations, and this corresponds to an *energy* density of 2–3 GeV per fm^3 – that is, enough pure energy present in the volume of a single proton to create three protons in line with Einstein's equation. How can physicists set about creating such extreme conditions?

The line of attack now being followed at CERN, in Europe, and at the Brookhaven National Laboratory in the United States, is to collide beams of heavy ions head on. These ions are the nuclei of heavy elements, from which the electrons have been stripped. So they each carry a large positive charge, made up of all the protons in each nucleus.

Scientists using particle accelerators routinely carry out experiments in which beams of protons or electrons (or their antimatter counterparts) are smashed into targets containing nuclei of heavier elements, or into opposing beams of elementary particles. But now researchers are developing the technique required to take beams containing nuclei of very heavy elements and smash them into opposing beams containing the same types of heavy nucleus.

To get a picture of the kind of collision that will result when two such nuclei meet head on, consider the (as yet hypothetical) example of what happens to a gold nucleus accelerated to 0.999 957 times the speed of light.

A gold nucleus contains 118 neutrons and 79 protons; so it has 79 units of positive charge, providing the handle by which it may be accelerated to such speeds using magnetic fields. At this speed, relativistic effects will make the mass of the nucleus increase, while it shrinks in the direction of motion to become a flattened pancake. These effects are predicted by Einstein's special theory of relativity, and have been routinely confirmed in countless experiments of this kind. A moving object really does get heavier and shrink in the direction of its motion as its speed gets closer to the speed of light.

The two effects involve the same relativistic factor, so in this particular example the mass increases to 108 times its rest mass, while the thickness of the nucleus along the line of flight shrinks to 1/108 times the thickness of a stationary gold nucleus. Do not worry about the odd 8. In round terms, it is a hundred times heavier (with a mass of more than 100 GeV per nucleon). But at the same time the thickness of the pancake has now shrunk to one hundredth of its former size – the thickness along the line of flight is now only 1 per cent of its diameter measured across the line of flight. With 100 times as much mass in one hundredth of the volume, the nucleus literally has a density 10,000 times greater (100 × 100) than the same nucleus at rest.

If such a relativistic nucleus meets an identical nucleus travelling the opposite way, the results will be spectacular. With (just over) 100 times as much mass in 1 per cent of the original volume for each nucleus, the density of matter in the colliding nuclei at the moment of overlap is more than 20,000 times the density of an ordinary gold nucleus; the same kind of density would be achieved in collisions between nuclei of other heavy elements, such as lead or uranium. And as the two nuclear pancakes try to pass through each other, there will be repeated collisions between protons and neutrons meeting head on, and between nucleons and the wreckage produced by collisions that have taken place just in front of them. The best picture of what happens then comes from physicists' standard model of nucleons as composed of quarks.

Each proton and neutron – each nucleon – contains three quarks. But, as I have mentioned, quarks cannot exist in isolation. They come either in triplets or in pairs, and the best way to understand this is to think of them as being held together by a piece of elastic (in reality, an exchange of gluons) holding two or three quarks together. This is, in fact, an example of the strong force at work. If you tried to separate two quarks, this elastic would stretch, and energy put into separating the two quarks would be stored in a way analogous to the way in which energy is stored in a stretched piece of elastic, or a stretched spring.

Up to a point, this means that two quarks joined in this way are held together more tightly the farther apart they are – the opposite of the way in which familiar forces like magnetism or gravity operate. Eventually, the stretched 'elastic' will snap, but only when enough energy has been put into the system to create two 'new' quarks ($E = mc^2$ again), one on each side of the break.

The process is reminiscent of trying to separate a north magnetic pole from a south magnetic pole by sawing a bar magnet in half. Every time you break the two poles apart, you find you are left with two new bar magnets, each with a north pole and a south pole, instead of two separated poles.

So the picture of the collision between two heavy ions moving at relativistic speeds is one in which quarks are ripped out of individual nucleons, stretching the elastic joining them to other quarks until it breaks, creating new combinations of pairs and triplets of quarks, with an overlapping tangle of breaking and rejoining elastic – like high-energy spaghetti. Tangled elastic may end up joining two quarks moving in opposite directions at close to the speed of light, absorbing large amounts of the kinetic energy of the collision and snapping to produce a host of new particles at the site of the collision, after what is left of the original nuclei has moved away.

This is the quark–gluon plasma that physicists are eager to study – the 'little bangs' in which conditions that may not have existed for 15 billion years, since the Big Bang itself, can be reproduced. And because particles are being manufactured out of energy (the relativistic kinetic energy of the colliding nuclei), it is easy to produce a mass of particles from this mini-fireball that is greater than the mass of the two original nuclei. Such particle collisions are *not* simply a question of breaking apart the incoming nuclei to release their constituent components, but are a means to create the high energy densities out of which new particles can be formed. The energy required to make the new particles has come from the magnetic fields used to accelerate the original nuclei.

How close are the experimenters to achieving the quark–gluon plasma? Existing particle accelerators were not built to do this kind of experiment, and quite apart from the energy input required there are other constraints which limit the kind of nuclei that can be used. At CERN, for example, the SPS accelerator can be operated only with nuclei that have equal numbers of protons and neutrons, while very heavy nuclei always have many more neutrons than protons. Working with nuclei of sulphur-32, the SPS can reach 19 GeV per nucleon, one-fifth of the way to the kind of collisions discussed here. Existing accelerators at Brookhaven can reach 5 GeV per nucleon, using silicon-28. With new booster systems (sometimes known as 'pre-accelerators'), both laboratories will soon be able to handle heavier nuclei, including lead, but only at about the same relativistic factors.

But in 1997 or 1998 Brookhaven's Relativistic Heavy Ion Collider (RHIC) and CERN's Large Hadron Collider (LHC) should both become operational and will operate with energy densities, temperatures and collision speeds well within the range in which, theory says, the quark–gluon plasma should form.

The grand unified theories were developed in large measure on the basis of results from colliding particle experiments, and will be tested further by the new generation of colliding ion experiments. They tell us that at very high temperatures the differences in strength between the forces disappear. The strong force, the weak force and electromagnetism would all be as strong as each other (gravity is the odd one out, and still cannot easily be included in these unified theories). The temperature at which this happens is the temperature of the Universe when it was 10^{-35} of a second old. Alan Guth, the father of the theory of inflation, explained how the quantum processes associated with the splitting of the fundamental forces at that time could have provided the boost (the more exotic kind of phase transition mentioned in Chapter Two) which inflated the Universe wildly at that time, before it settled down to the more sedate expansion that we see today.

What this means is that instead of the Universe having expanded from a region one millimetre across which contained 10^{24} separate domains, just one of those domains first expanded from 10^{-25} cm – not just up to a millimetre across, but up to about a kilometre – all in about 10^{-32} of a second. To put this in perspective, it is equivalent to inflating a tennis ball to the size of the observable Universe today, all in just 10^{-32} of a second. At that point inflation ended, and the very smooth and uniform early Universe expanded on its way, passing through the quark–hadron phase transition when it was a hundred-thousandth (10^{-5}) of a second old. Our entire observable Universe comes from a region about a millimetre across *within* that primordial region of inflated super-smoothness a kilometre or so across. The reason why COBE found such a smooth background radiation is that inflation means that the entire observable Universe has developed from a small part of a single original domain.

Although inflation was a huge success at explaining the uniformity of the Universe, in the 1980s it seemed to suffer from two difficulties. The first is that although it had been very successful at explaining things we already knew, it had never made a successful prediction. To emphasize the importance of genuine *pre*dictions, scientists sometimes call an explanation made after the event a '*post*diction'. Inflation's great success in explaining why the Universe is uniform is a postdiction – we knew the Universe was uniform, and Guth explained why. But scientists only really trust theories that predict things we did not know at first, but which are later proved to be right by observations and experiment – genuine *pre*dictions.

When the cosmologists looked carefully at the details of how inflation ought to work, they found, rather to their surprise, that the process is not *perfectly* uniform after all. Even a domain as ridiculously small as 10^{-25} cm across is still big enough for quantum fluctuations to go on inside it, producing tiny ripples in its structure. The theory said that inflation should leave behind an expanded version of these fluctuations, some very small irregularities in the Universe – tiny ripples in the distribution of matter and energy. What is more, the required ripples were predicted to have a particular pattern, with the same pattern showing up on all scales.

The ripples in the background radiation found by COBE exactly match this predicted pattern of ripples left over from the era of inflation. This was one of the major reasons why cosmologists were so excited when the discovery of those ripples was announced – the inflationary model had made its first successful prediction, placing it on a more secure footing than before. And because the imprint of those ripples was stamped on the Universe in the era of inflation, when the Universe was still less than 10^{-30} of a second old, it means that the ripples in time observed by COBE are actually giving us information about the Universe not just when it was 300,000 years old, but from the first 10^{-30} of a second of its existence.

This is also a triumph for the grand unified theories, as well as for inflation. Inflation, after all, is based on the grand unified theories. The success of this whole package of ideas in explaining the very early history of the Universe and making predictions that have now been tested and have passed those tests is the best evidence we have that the grand unified theories themselves are correct. Particle physicists, interested in the world of the very small, now find that the best way to test their theories is by looking at the entire Universe and its history right back to the beginning. Indeed, one reason why physicists can now confidently predict what they expect to find when gold nuclei are collided at nearly the speed of light is that in a very real sense their theories have already been tested under the more extreme conditions of the Big Bang itself. The ripples from 15 billion years ago found by COBE help to tell us what happens to colliding gold nuclei at Brookhaven in the 1990s. The whole package hangs together breathtakingly well, and implies that fundamental science really does reveal deep truths about the nature of the Universe.

This is worth emphasizing, because even today some people try to dismiss the whole scientific process as either misguided or incorrect. Such individuals may say that they 'cannot believe in', say, the general theory of relativity, or quantum physics, because it runs counter to common sense. Or they pooh-pooh the notion that human beings could ever really understand what goes on inside atoms, or in the Big Bang. But all of science is cut from the same cloth. The same theories that explain how electricity works, or the fall of an apple from a tree, explain how particles behave in 'atom smashing' experiments, or in the first split second of the Big Bang. You cannot remove one piece you do not like (such as the theory of relativity) without unravelling the whole cloth. And the fact that the cloth does not unravel, even under the most extreme conditions we are able to probe – the ripples from the first 10^{-30} of a second revealed by COBE – shows just how strong it is. If you think you can come up with a better theory than Einstein's and the grand unified theories, it had better be able to explain *everything*, including those ripples, or it has no hope of being taken seriously.

The second difficulty that cosmologists had with the notion of inflation in the 1980s was that, in its original form, Guth's theory seemed to require the details of the inflation process to be rather precisely 'fine tuned' to produce the kind of Universe we live in. This is very reminiscent of the way in which the strengths of the forces of physics today, and the masses of particles such as protons and neutrons, seem to be rather finely tuned to allow for the existence of stars, spiral galaxies, black holes, complex heavy elements and ourselves. At first sight, there is no obvious reason why the inflation process should have gone on for just long enough and at just the right rate to produce a Universe in which stars and galaxies can form. A shorter, less intense burst of inflation would have left the proto-Universe too jumbled up, and also in danger of quickly recollapsing all the way back down into a singularity; a longer, stronger burst of inflation would have spread the stuff of the proto-Universe so thin that no stars and galaxies could ever form.

This fine-tuning problem is generally regarded as the biggest difficulty with inflation, and there have been many attempts to get rid of the problem, all of which involve making the theory more complicated. But I believe this line of approach to be wrong. The problem is simply another example of the Goldilocks effect – why is inflation, like so many other properties of the Universe, 'just right' to allow our existence? The simplest version of inflation, complete with the fine-tuning problem, can explain everything we can see, including the ripples in the cosmic background radiation. And the fine-tuning problem itself can be resolved in the same way that we can resolve the puzzle of the fine-tuning of the forces of physics, once we accept that the Universe itself is alive and has evolved.

Before going into details of the evolution of the living Universe, though, there is one last loose end to be cleared up. By October 1992, the analysis of the COBE data had slotted one last piece into the cosmic jigsaw puzzle. Having confirmed the accuracy of the inflationary scenario and the Big Bang model, and having already established beyond doubt that there must be a hundred times more dark matter in the Universe than the matter we can see in the form of bright stars and galaxies, at last COBE revealed what the dark matter is.

Although speculation that there might be more to the Universe than meets the eye goes back more than half a century, it was only in the 1980s that the mounting weight of evidence began to convince most astronomers. Studies of how individual galaxies rotate, and how groups of galaxies move together through space, provided compelling evidence that there must be a lot of dark stuff around. But how much? The reason it took a long time for astronomers to accept that there might really be enough dark stuff around to make the Universe flat (just closed) was, as much as anything, prejudice. They were bright-stuff chauvinists, and found it hard to come to terms with the idea that generations of astronomers had devoted their lives to studying what might turn out to be no more than 1 per cent of the Universe.

As recently as 1981, I asked an eminent astronomer, John Huchra of the Smithsonian Institution Observatory in Massachusetts, for his views on dark matter. His reply is typical of the way astronomers thought at the time. He acknowledged that all the evidence pointed to the existence of a large amount of dark matter in the Universe, but went on to say that 'my own interpretation of the existing data is that the Universe is open, but only by a factor of three to five. From a philosophical point of view, an optical observer would quickly lose interest in observational cosmology if the Universe was dominated by things he couldn't see.'

This may have been typical of the way astronomers thought at the time, but Huchra was, like virtually all his colleagues in the early 1980s, wrong on both counts. We now know that the Universe is indeed just closed, not open. And yet, optical observers (people who study the Universe using ordinary light) have found plenty to interest them and to study, even though the Universe is dominated gravitationally by things they cannot see. What Huchra has not realized in 1981 is that although 99 per cent of the Universe may be in the form of dark stuff, the dark stuff reveals not just its presence but a great deal more about itself through its gravitational influence on the bright stuff. It interacts with visible matter in such specific ways that studies of the bright stuff can tell us a great deal about the things we cannot see. As a result, proof that the dark stuff is there meant more work, not less, for observational cosmologists (including Huchra himself).

The bright galaxies act as 'tracers' of the dark matter. In one analogy, they can be likened to snow on the top of a mountain range. Even if you could not see the rocks of which the mountains were made, but could only observe the snow, you would still know that the mountains were there, and you could still infer a great deal about the nature of mountains from studying the patterns made by the snow. Or you can think of the bright galaxies as like foam on the surface of the sea, only existing because of the heaving mass of water below. By studying the patterns made by the foam, you could work out a great deal about the nature of water and how waves move. All of the ingenuity of observational cosmologists is now applied to understanding what kind of waves produce the foam of galaxies that we see in our Universe today.

The clinching evidence that established once and for all that the Universe really is just closed was presented to a meeting of The Royal Society in London, in November 1985. It came from a survey of the distribution of galaxies across the sky carried out by sensors on board a satellite known as IRAS, the Infrared Astronomical Satellite.

All studies of the distribution of galaxies seen in visible light are handicapped by the phenomenon known as reddening. This has nothing to do with the redshift, remember, but is essentially the same as the way dust in the atmosphere of the Earth makes sunsets red. The dust scatters light with shorter (which means bluer) wavelengths, and leaves the red more or less alone. If there is enough dust, there is only a dim red light left to see. And the same thing happens with dust in the Milky Way, blocking out the light from many parts of the sky. Light from faint galaxies (which, by and large, means more distant galaxies) is more severely affected. But red light is not affected very much, and infrared radiation (a form of light invisible to our eyes, with wavelengths even longer than red light) is affected even less. So telescopes and detectors sensitive to infrared light can probe much deeper into the Universe when making surveys of galaxies.

The trouble is that infrared light is blocked by water vapour in the Earth's atmosphere. There are infrared telescopes in use on the tops of high mountains, above most of the water vapour. But to get a really good infrared snapshot of the Universe the detectors need to be hoisted into orbit, above the atmosphere entirely. The IRAS detectors have provided the best in-depth picture of the number and distribution of galaxies yet available, and although the picture has been updated and improved since 1985, the conclusion is still the same. It is based on a comparison of this infrared picture of the Universe with what was already known about the microwave background in the mid-1980s.

Although I have stressed the extreme uniformity of the microwave background, there is actually one way in which it is different in different parts of the sky. There is a slightly warmer patch in the sky (corresponding to a tiny blueshift in the background radiation) in one direction, and a slightly cooler patch (corresponding to a tiny redshift) in the opposite direction. The natural explanation of this is that our entire Local Group of galaxies is moving through the background radiation (which means moving relative to the expansion of the Universe) at a speed of about 600 km per second.

This is an interesting discovery in its own right, and since there are many galaxies moving together it is known as a 'streaming' motion. For a time, astronomers puzzled over why nearby galaxies, and the Milky Way, should be streaming in the same direction. But IRAS found the answer. The infrared survey shows that there is a concentration of galaxies in the sky in exactly the direction of this streaming motion. Where we see the 'snow' of galaxies, of course, there must also be all the mass of the 'mountains' of dark matter associated with those galaxies. The explanation of the streaming motion is, after all, simple: we are moving in that direction at 600 km per second because there is an extra concentration of matter in that direction, tugging us with its gravitational attraction.

Michael Rowan-Robinson and colleagues at Queen Mary and Westfield College, London, worked out how much matter there must be in the Universe if the dark stuff is distributed in roughly the same way as the bright galaxies, and the extra concentration of matter in the direction we are moving is exactly enough to produce streaming at a rate of 600 km per second. You will not be surprised to learn that the overall density of matter required is exactly what we need to make the Universe flat, or just closed. This is the most powerful piece of evidence from studies of the distribution of galaxies that 99 per cent of the Universe is in the form of dark matter – but it came, as John Huchra was no doubt relieved to note, from studies of ordinary, optically visible galaxies (stretching a point just a tiny bit to include infrared radiation under the heading of 'visible' light).

In 1992, that same Michael Rowan-Robinson was one of a handful of researchers who hurried to interpret COBE's discovery of ripples in the background radiation within the context of a Universe dominated by dark matter. By then, it seemed that there were two kinds of dark matter to choose from, and during the second half of the 1980s there had been considerable rivalry between proponents of the two different schools of thought.

Once astronomers had come to terms with the fact that 99 per cent of the Universe must be in a form that could never be seen, the first and most natural assumption was that the dark stuff must be something that we already knew about. There was only one such possible candidate – the particles known as neutrinos, that we have already met as key players in the supernova saga.

The neutrino is surely the most bizarre particle yet identified, even though particle physicists now know of the existence of literally hundreds of 'fundamental' particles, and feel no qualms about invoking particles yet to be discovered to explain the behaviour of the subatomic world. The physicists manufacture particles out of pure energy in proliferation, as we have seen, by smashing beams of protons or electrons into one another at accelerator laboratories such as CERN and Brookhaven. Yet it is less than a hundred years since J. J. Thomson, working at the Cavendish Laboratory in 1897, established that 'cathode rays' are actually particles, carrying negative electric charge, that can somehow be chipped away from the atom, once thought to be indivisible. And it is scarcely sixty years – less than a human lifetime – since it was first appreciated that there is even more to the particle world than these electrons and the positively charged protons which reside in the nuclei of atoms.

The basics of our modern understanding of the atom as a tiny, positively charged nucleus surrounded by a cloud of negatively charged electrons was established by Ernest Rutherford (later Lord Rutherford), then working at the University of Manchester, as recently as 1911. In the experiments which led Rutherford to develop this model of the atom, beams of positively charged alpha particles were fired at thin metal foils. Most of the particles went straight through the foil, unimpeded; just a few, though, bounced back in the direction they had come from. Rutherford's model explained these observations, because most of the alpha particles brush through the electron clouds surrounding the atomic nuclei in the foil, and only those that happen to collide almost head on with a nucleus are repelled by the positive charge it carries with sufficient strength to 'bounce'.

Statistical analysis of these experiments showed that the nucleus is typically only about one-hundred-thousandth the size of an atom – a nucleus about 10^{-13} cm across embedded in an electron cloud some 10^{-8} cm across. The alpha particles that proved such a useful probe of atomic structure had been investigated by Rutherford (who gave them their name) and his colleague Frederick Soddy (later Sir Frederick), then at McGill University, Montreal, in the early years of the twentieth century. Their studies showed that radioactive atoms could emit two kinds of 'ray' – alpha-rays (alpha particles under a different name), which we now know to be the nuclei of helium atoms, and beta-rays, later identified as electrons. When a third type of radiation was later discovered, it was, of course, called gamma radiation. Gamma-rays are actually a form of high-energy electromagnetic radiation, similar to X-rays.

For twenty years after Rutherford came up with his nuclear model of the atom, nobody seriously questioned the idea that physicists had identified all the important particles of nature. There were just two of them, electrons and protons. Because alpha particles (helium nuclei), for example, have the mass of four protons and a charge just twice that of a proton, it was thought that some electrons (in this case, two) could also be bound up in atomic nuclei – and since opposite charges attract one another, there seemed no reason to think this at all unlikely.

Indeed, over the fifteen years or so following the development of Rutherford's model, physicists were much more concerned with explaining why *all* the electrons in the atomic cloud did not fall into the nucleus. The explanation required the development of quantum theory, which was essentially complete by 1926 (see *In Search of Schrödinger's Cat*). Only then, in the second half of the 1920s, did physicists begin to worry seriously about a puzzling feature of beta radiation.

In beta 'decay', as it is called, atoms eject electrons. These electrons do not come from the cloud around the nucleus, so they must come from the nucleus itself. But here lay the puzzle. If an electron shoots out of a nucleus at high velocity, there ought to be a discernible recoil in the nucleus itself – like the kick of a rifle being fired. But physicists in the 1920s could find no evidence for any such recoil, and for a time they seriously considered the possibility that the laws of conservation of energy and momentum might not work in atomic nuclei. An alternative explanation, seemingly no less desperate at the time, came from an Austrian-born physicist working in Zurich.

Wolfgang Pauli, born in Vienna in 1900, was known for his clear thinking. He had made his name as a nineteen-year-old student by producing what was then the clearest account of Einstein's two theories of relativity. He saw how to cut through the Gordian knot of the beta decay problem, and in a letter to Lise Meitner (one of the physicists whose work led to an understanding of nuclear fission) he made the straightforward proposal that the 'extra' momentum required to balance the books was being carried off by another particle which must be emitted from the nucleus at the same time as the electron observed in beta decay.

Only, the proposal was not that straightforward. Such a particle, unobservable by the technology of the day, might never be detected, thought Pauli, but it had to be there for the conservation laws to be obeyed. In order to be undetectable, this hypothetical particle would have to have no electric charge (which seemed absurd, since every particle known at the time carried charge), and essentially zero mass (equally absurd). Its only property would be quantum spin – a bizarre property in its own right, since quantum entities have to rotate *twice* to get back to where they started.

This strange package of ideas to 'explain' beta decay, published formally in 1931, did not meet with instant acclaim. It seemed both too far-fetched and too easy, a cop-out holding out the threat of invoking a new kind of undetectable particle to explain every puzzling phenomenon in experimental physics. The notion made so little impact, indeed, that the name Pauli had suggested for his hypothetical particle, the 'neutron', was hijacked a year later and applied to a newly discovered particle with roughly the same mass as the proton.

This discovery was dramatic enough to be going on with. Atomic nuclei were composed not of a mixture of protons and electrons, but of protons and neutrons (two of each in the alpha particle). And, it turned out, in beta decay a neutron was transformed into a proton and an electron. But there was still the problem of missing momentum associated with beta decay, and Pauli persisted in promoting the idea of yet another neutral particle. Eventually, in 1933, he won support from an Italian-born physicist a year his junior, Enrico Fermi. Fermi took up Pauli's idea and put it on a more respectable footing by introducing a new force into the calculations to go with the new particle – the weak nuclear force.

Fermi modelled his description of the new force on the explanation of electric force in terms of the exchange of photons (the particles of light) between charged particles. He suggested that in beta decay a neutron actually emits an entity rather like a photon, but carrying one unit of negative charge. In the process, the neutron becomes a proton. The photon-like charged particle that is ejected itself promptly decays into an electron and the hypothetical particle proposed by Pauli. With the name 'neutron' now spoken for, Fermi called this particle the 'neutrino' ('little neutral one').

The weak interaction described by Fermi is the only interaction that neutrinos deign to take part in, unless they do have a tiny mass and can feel the force of gravity. The weak force is so weak that if a beam of neutrinos travelled through solid lead for 3500 light years, only half of them would be absorbed along the way by the nuclei of the lead atoms. According to standard theory today, updating Fermi, about one-tenth as much energy as the Sun emits in visible light is emitted in the form of neutrinos, and billions of these ghostly particles are zipping through your body every second, without your body noticing them, or the neutrinos noticing your body. It is a measure of the extent to which the notion of neutrinos and the weak interaction failed to set the scientific world on fire that in 1933 the respected journal *Nature* rejected a paper from Fermi setting out these ideas as 'too speculative'. But his work was soon published in Italian, and not long after in English (though not in *Nature*).

The discovery of the neutron had broken the ice to some extent, demonstrating both that neutral particles could exist and that there was more to the atom than just protons and electrons. Further investigations of subatomic processes continued to show up the need for neutrinos if the conservation laws were to be obeyed, and the notion of these ghost-like particles gained acceptance, even though it seemed unlikely that they would ever be detected. Indeed, Pauli himself offered a case of champagne as a reward to any experimenter who successfully took up the challenge of detecting neutrinos, probably never anticipating that he would have to pay up. But in the 1950s a series of experiments carried out by Frederick Reines and Clyde Cowan, culminating in 1956, proved that neutrinos exist.

Reines and Cowan placed a tank containing 450 kg (1000 lb) of water alongside the Savannah River nuclear reactor in the United States. According to theory, a flood of neutrinos must be being produced by the nuclear reactions going on inside the reactor, and one or two of them ought to interact with atoms in the tank of water every hour.

The reaction that they looked for, in a series of tests they called Project Poltergeist, was actually the inverse of beta decay. In this reaction, a neutrino (actually an antineutrino) strikes a proton and converts it into a neutron, while a positron (the positively charged counterpart of the electron) carries away the positive charge. It was the positrons that the Savannah River experiment actually detected; Reines and Cowan sent Pauli a telegram informing him of their success, and Pauli duly made good his twenty-five-year-old offer by sending them a case of champagne.

Although the original idea was that neutrinos must have precisely zero mass, and no experiment has, in fact, ever measured the mass of a neutrino, there are so many of these particles filling the Universe that if each of them had even a very small mass, much less than the mass of an electron, the total would add up to the amount required to make the Universe closed. Equally, though, if neutrinos had anything more than a very tiny mass their combined gravitational influence would be enough to make space much more highly curved (and the Universe 'more closed') than we actually see.

Because neutrinos have very little mass, they are born, in beta decay and similar reactions, moving very fast – at the speed of light if they have zero mass; very close to the speed of light if they have a little mass. For this reason, they are known as 'hot' particles, or 'hot dark matter'. Hot dark matter particles emerging from the Big Bang would tend to break up any small-scale concentrations of atomic matter in the early Universe, scattering clumps of ordinary atomic stuff, 'much as a cannon-ball moving at high speed might scatter a loosely built wall of bricks without being appreciably slowed by the collision', in the graphic words of Jack Burns, of the University of New Mexico. As a result, the distinctive feature of the kind of distribution of matter that would emerge in a Universe dominated by hot dark matter is that large structures would form first, as the neutrinos cooled and slowed, and these huge, pancake-like structures would then break down into smaller structures, clusters of galaxies and individual galaxies, as the Universe continued to expand. This is known as a 'top-down' process.

Unfortunately for the proponents of the hot dark matter hypothesis, though, studies of the distribution of the galactic froth in the Universe today show that this top-down structure does not match the actual distribution of matter. Instead, the pattern of galaxies across the sky matches much more closely the kind of pattern that would be produced if small clumps of matter had got together first after the Big Bang, and then the small clumps had clumped together to make bigger clumps. Galaxies formed first, then clusters, then superclusters, and so on. This is known as the 'bottom-up' scenario.

Bottom-up structure would arise naturally in the Universe if the dark matter left over from the Big Bang was in a 'cold' form right from the beginning. Cold dark matter consists of particles that emerged from the Big Bang with little or no speed, and just sat around, participating in the expansion of the Universe and tugging on each other, and on atomic matter, gravitationally. But in order not to show up in other ways, this cold dark matter must also decline to interact with atomic matter in any other way except through gravity. A combination of large amounts of cold dark matter and the observed smattering of bright stuff can explain in broad outline how galaxies are scattered across the Universe today, with one obvious snag – nobody has ever detected a cold dark matter particle.

This is not too depressing, because there are other good reasons for suspecting that such particles must exist. In particular, they are required to exist by the best versions of grand unified theories, those very theories which have proved so successful in so many other ways, including explaining the Big Bang in terms of inflation. It would be a severe embarrassment to physics if there proved to be no cold dark matter particles, and the possible candidates required by the theory have even been given names, long before cosmologists decided that the Universe may be filled with such particles. The most likely contender is a particle known as the axion.

Like neutrinos, the hypothetical cold dark matter particles do not carry electric charge, and they do not feel the strong force, or (unlike neutrinos) even the weak force. All they feel is gravity. Each of these particles might have a mass rather greater than that of a proton, and there could be several dozen of them in each litre of air that you breathe. Rather more significantly, there could be several dozen of them in each litre of 'empty space' throughout the entire Universe. Together, they could close the Universe gravitationally; but individually they are extremely hard to detect, although several experiments are now under way around the world to try to track them down.

In the mid-1980s, hot dark matter (in the form of neutrinos) was the preferred choice for the dark stuff. In the late 1980s, cold dark matter (in the form of axions or something similar) seemed a better bet. But then the picture became more complicated. In the early 1990s, cold dark matter scenarios were investigated in more detail by comparing computer simulations of the distribution of galaxies in a cold dark matter Universe with the patterns seen in the sky. The simple cold dark matter picture began to run into difficulties. The pattern of bright galaxies in the sky has a little too much structure on the large scale, and seems to require an additional influence to be at work, as well as the cold dark matter itself. Cold dark matter *nearly* fits the bill, but not quite.

One possible resolution of the puzzle, provided by COBE, is disarmingly simple. Because the pattern of ripples found by COBE is the same on all scales, it is reasonable to extrapolate this down to scales smaller than the resolution yet achieved by COBE's sensors themselves, filling in the picture of how matter is distributed on the scale of individual clusters of galaxies, clusters of clusters, and so on. This in turn matches up with the observed distribution of actual galaxies in the Universe today. In order to explain the kinds of structure revealed by COBE, and to match the pattern of galaxies in the sky today, what you need is a mixture of about two-thirds cold dark matter, one-third hot dark matter and just a smear of ordinary atomic matter. In such a 'mixed dark matter' scenario, the cold dark matter provides the clumps on which galaxies and clusters of galaxies grow, while the hot dark matter fills in some of the space in between the cold dark matter clumps, smoothing out the overall density of the Universe and reducing the contrast between the clumps and the spaces. Atomic matter – the bright stuff of stars and galaxies – feels the gravitational influence of both kinds of dark matter, and so the foam we see today represents the averaged-out influence of waves made up of hot and cold dark matter.

Several groups hit on this realization almost simultaneously, within weeks of the announcement of the COBE results. The group at Queen Mary and Westfield College, including Michael Rowan-Robinson, gave the most precise estimate. Pointing out the need to match the new discoveries with the IRAS observations which show the Universe to be flat, they said that 69 per cent of the mass required must be in the form of cold dark matter, 30 per cent in the form of hot dark matter, and just 1 per cent in the form of atomic stuff, the stuff of which we and the stars are made.

The precise figures should, perhaps, be taken with just a pinch of salt. The more rough and ready 'two-thirds cold, one-third hot' dark matter is probably a more realistic description of our state of knowledge about the make-up of the Universe. And there is one other reason for caution – it is still possible to explain the deviations from the 'pure' cold dark matter scenario another way, by invoking gravitational radiation.

Gravity waves are literally ripples in the structure of spacetime itself, and they should have been produced during the inflationary era of the birth of the Universe. Some calculations suggest that the influence of these waves rippling through the Universe would be just right to explain the discrepancies between the pure cold dark matter model and the observations, without invoking hot dark matter at all. We should soon know for sure, because although the influence of gravity waves cannot be seen directly by COBE, it should, if it is there, show up as a direct influence on the background radiation at smaller angular scales, which are now being probed by ground-based detectors.

If you take the Rowan-Robinson team's calculations at face value, however, with this ratio they can even tell us how much mass each individual neutrino must have, if neutrinos are indeed the hot dark matter particles. It works out at just 7.5 electronvolts. The electron, which is the lightest particle that has any real direct influence on our daily lives, has a mass of about 500,000 eV, equivalent to 10^{-30} kg. So the mass of the neutrino is about 0.0014 per cent of the mass of the electron, and it is no surprise that nobody has yet been able to measure it. The surprise, in fact, is that several attempts have been made to measure the mass of the neutrino, and that they all agree that it must be less than about 20 eV. In other words, those experiments would have recorded a mass had it been bigger than 20 eV, and since they did not record a mass it must be less than that. It could be precisely zero; or it could very well be, as some cosmologists are now saying, 7 or 8 eV. The experimenters are tantalizingly close to being able to measure such a small mass, and if and when they do it will be the greatest triumph yet of the complete cosmological package based on inflation, the Big Bang and dark matter.

All of this further allays John Huchra's fear, in 1981, that there would be no point in observational cosmology if 99 per cent of the Universe were in the form of dark matter; observational cosmology, with no information except that provided by the electromagnetic radiation from 1 per cent of the stuff of the Universe, can tell particle physicists how much mass the neutrino has, even though particle physcists themselves cannot yet measure that small a mass!

It seems that we know, more precisely than anybody has ever known before, what the Universe is made of, and how much of the different kinds of stuff there are, as well as how the Universe came into existence. We know that it seems to be so efficient at the job of making stars and turning them into black holes that it could almost have been designed for the job. And we know that the ultimate fate of the Universe itself is that one day the present expansion will be first halted and then reversed, so that it collapses back into a singularity that is a mirror-image of the one that gave it birth. We actually live inside a huge black hole – a black hole so big that it contains billions of other black holes inside itself. And what is more, we have a pretty good idea of what happens to anything that collapses towards a singularity inside a black hole.

When mathematical physicists first began to think about black holes and singularities, they did not worry too much about what happened to the matter that fell into a singularity. First, since singularities only seemed to occur inside black holes, where they could not be seen, it did not seem to matter much what happened to them. And then, since the laws of physics seemed to break down at a singularity, most of the researchers seemed content to say that matter was literally squeezed out of existence at such a point of infinite compression.

But the realization that our own Universe seems to have been born out of a singularity, and the evidence, confirmed by COBE, that we live inside a black hole, pulls the rug from under the argument that we need not concern ourselves with what goes on inside a black hole. We certainly do want to know what goes on inside our Universe!

The idea of the Universe as a black hole is not new, although until recently it was distinctly unfashionable. As far as I know, I was the first person to describe the Universe in these words, in an unsigned editorial commentary in the journal *Nature* in 1971 (volume 232, page 440). Scarcely anybody took the notion seriously, because nobody then realized that the Universe is dominated gravitationally by dark matter. But today, scarcely anybody doubts this picture. And if all the complexity of galaxies, stars, planets and organic life has emerged from the singularity in which our Universe was born, within a black hole, could not something similar be happening to the singularities at the hearts of other black holes?

The most naive expectation of what might happen to a collapsing singularity to turn it into the kind of expansion from a singularity that we see in our Universe is that there is simply a 'bounce' at the singularity, turning collapse into expansion.

Unfortunately, that will not work. A singularity forming from a collapse within our three dimensions of space and one of time cannot turn itself around and explode back outwards in the same three dimensions of space and one of time. But, in the 1980s relativists realized that there is nothing to stop the material that falls into a singularity in our three dimensions of space and one of time from being shunted through a kind of spacetime warp and emerging as an expanding singularity in another set of dimensions – another spacetime.

Mathematically, this 'new' spacetime is represented by a set of four dimensions (three of space and one of time), just like our own, but with *all* the new dimensions at right angles to *all* the familiar dimensions of our own spacetime. Every singularity, in this picture, has its own set of spacetime dimensions, forming a bubble universe within the framework of some 'super' spacetime, which we can refer to simply as 'superspace'.

One way to picture this is to go back to the analogy between the three dimensions of expanding space around us and the two-dimensional expanding surface of a balloon that is being steadily filled with air. The analogy is not with the volume of air inside the balloon, but with the expanding skin of the balloon, stretching uniformly in two dimensions but curved around upon itself in a closed surface. Imagine a black hole as forming from a tiny pimple on the surface of the balloon, a small piece of the stretching rubber that gets pinched off, and starts to expand in its own right. There is a new bubble, attached to the original balloon by a tiny, narrow throat – the black hole. And this new bubble can expand away happily in its own right, to become as big as the original balloon, or even bigger, without the skin of the original balloon (the original universe) being affected at all. There can be many bubbles growing out of the skin (the spacetime) of the original universe in this way at the same time. And, of course, new bubbles can grow out of the skin of each new universe, *ad infinitum*.

Instead of the collapse of a black hole representing a one-way journey to nowhere, many researchers now believe that it is a one-way journey to somewhere – to a new expanding universe in its own set of dimensions. Instead of a black-hole singularity 'bouncing' to become an exploding outpouring of energy blasting back into our Universe, it is shunted sideways in spacetime.

The dramatic implication is that many – perhaps all – of the black holes that form in our Universe may be the seeds of new universes. And, of course, *our own Universe may have been born in this way out of a black hole in another universe*. While the fact that the laws of physics in our Universe seem to be rather precisely 'fine tuned' to encourage the formation of black holes means that they are actually fine tuned for the production of *more universes.*

This is a spectacular shift of viewpoint, and most cosmologists are still struggling to come to grips with it. If one universe exists, then it seems that there must be many – very many, perhaps even an infinite number of universes. Our Universe has to be seen as just one component of a vast array of universes, a self-reproducing system connected only by the 'tunnels' through spacetime (perhaps better regarded as cosmic umbilical cords) that join a 'baby' universe to its 'parent'. It is relatively easy to see how such a family of universes can continue to exist, and to reproduce, once something like our own Universe exists. But how did the whole thing get started? Where did the first universe, or universes, come from?

The best answer seems to come from one of the stranger implications of quantum theory. And it is scarcely less strange to learn that it was first proposed, in a simple form, in the early 1970s, before ideas such as inflation had been dreamed of, and before the COBE proposal was even a gleam in John Mather's eye.

The key concept is quantum uncertainty. This says that there is always an intrinsic uncertainty in many physical properties of the Universe and things in the Universe. The most commonly quoted example is the uncertainty that relates the position of a particle to its motion. Momentum is a measure of where a particle is going, and quantum uncertainty makes it impossible to measure the position of, say, an electron *and* its momentum *at the same time.*

This is not a result of the inadequacies of our measuring equipment, but a fundamental law of nature which has been thoroughly tested and proved in many experiments. An object like an electron simply does not have both a precise momentum and a precise position.

This is related to the fact that in the quantum world 'particles' are also waves. Waves, like ripples on a pond, tend to have a pretty clear direction, but they are spread-out things which do not have a definite location. Particles are more easy to pin down in terms of position, but they do not have the same in-built sense of direction that a wave has. So any entity that shares properties we usually attribute to waves and properties we usually attribute to particles will be a little uncertain both about where it is and where it is going.

But such effects show up only on the very small scale. The archetypal 'wavicle', showing both particle properties and wave properties, is the photon, the 'particle' of light. Electrons also show this dual personality, but it scarcely shows for anything bigger than an atom, and is of no practical consequence when we are dealing with waves on the ocean, or 'particles' the size of a tennis ball.

Another pair of uncertain variables linked in this way are energy and time. Again, the uncertainty applies only on the subatomic scale, as far as any practical consequences are concerned. But what quantum physics tells us is that any tiny region of the vacuum, which we think of as 'empty' space, might actually contain a small amount of energy for a short time. In a sense, it is allowed to possess this energy if the Universe does not have time to 'notice' the discrepancy. The more energy is involved, the shorter the time allowed. But because particles are made of energy ($E = mc^2$), this means that particles are allowed to pop into existence in the vacuum of empty space. They are made out of nothing at all, and can exist only if they pop back out of existence again very quickly.

In this picture, the quantum vacuum is a seething froth of particles, constantly appearing and disappearing, and giving 'nothing at all' a rich quantum structure. The rapidly appearing and disappearing particles are known as virtual particles, and are said to be produced by quantum fluctuations of the vacuum.

It may seem that quantum theory has run wild when pushed to such extremes, and common sense might tell you that the idea is too crazy to be true. Unfortunately for common sense, these quantum fluctuations have a measurable influence on the way 'real' particles behave. The nature of the electric force between charged particles, for example, is altered by the presence of virtual particles, and measurements of the nature of the electric force show that it matches the predictions of quantum theory, rather than the common-sense way it would behave in a 'bare' vacuum.

In one of the strangest and least publicized fundamental experiments in physics, the consequences of vacuum fluctuations can actually be measured directly. The easiest virtual particles to make in this way are the particles of light, photons. After all, photons have zero mass, and so the only energy you need to 'borrow' from the vacuum is the energy of the electromagnetic waves associated with the particles. But the nature of the electromagnetic waves that can exist in the vacuum depends on their surroundings.

Way out in empty space, between the stars, all kinds of waves, with any wavelength imaginable, can pop in and out of existence. But you can inhibit the process of vacuum fluctuations simply by placing two ordinary metal plates close together, with a thin gap between the faces of the plates. Between two metal plates, electromagnetic waves can form certain stable patterns only. The waves that bounce from one plate to the other behave rather like the waves on a plucked guitar string. You have to have a whole number of wavelengths to fit the gap between the plates, just as you can play only those notes on a guitar string which have a wavelength matching the length of the string. As a result, although the vacuum still fluctuates in the gap between the plates, and still produces virtual photons out of nothing at all, it cannot produce as many photons in the gap as it does outside.

Switching back from a wave description of electromagnetic phenomena to the particle description, in each cubic centimetre of vacuum there are fewer photons bouncing around between the plates than there are outside. The result is a pressure which tries to squeeze the plates together, a force of attraction between the two plates. The force, known as the Casimir force after the Dutch physicist who predicted its existence in 1948, has actually been measured. The implication is that the vacuum really is seething with virtual particles and short-lived bursts of energy.

What has all this got to do with the creation of universes? It all hinges on the fact that we live inside a black hole. Back in the early 1980s, R. K. Pathria, of the University of Waterloo, Ontario, used the suggestion that our Universe might be a black hole as the jumping-off point for some proper calculations, within the framework of the general theory of relativity, that put the notion on a more secure scientific footing (*Nature*, volume 240, page 298). Then, in 1973, a paper from Edward Tryon of the City University of New York took the argument a step further, suggesting that our entire Universe might simply be a fluctuation of the vacuum. Hardly surprisingly, the paper duly appeared in *Nature* (volume 246, page 396), where he pointed out the curious fact that our Universe contains *zero* energy – provided it is indeed closed and forms a black hole. The point Tryon jumped off from – the secret of making universes out of nothing at all, as vacuum fluctuations – is that the gravitational energy of the Universe is negative.

Think of it this way. For a collection of matter, such as the atoms that make up a star, or the bricks that make up a pile, the 'zero of gravitational energy' associated with those objects is when they are far apart – as far apart as it is possible for them to be. The strange thing is, as the objects fall together under the influence of gravity they *lose* energy. They start with none, and end up with less. So gravitational energy is negative, from the perspective in which everyday energy (the mc^2 in those atoms and bricks) is positive. Any object in the Universe, like a planet or the Sun, which is *not* spread out as far as possible literally has a negative amount of gravitational energy. And if it shrinks, its gravitational energy becomes more negative.

The reason why this was so interesting to Tryon is that the energy of all the matter in the Universe, all the mc^2, is positive. What is more, if you take a lump of matter and squeeze it into a singularity, then at the singularity the negative gravitational energy of the mass is exactly equal and opposite to its mass energy.

If this blows your mind, you are in good company. One of my favourite Albert Einstein anecdotes recounts a tale from half a century ago, during the Second World War, when Einstein acted as a part-time consultant to the US Navy, assessing ideas for new weapons. Einstein did not actually work in Washington, but every couple of weeks George Gamow, another eminent physicist, would bring a briefcase full of ideas up to Princeton for Einstein to peruse. One day, as Gamow recalled in his book *My World Line*, the two physicists were crossing the road together, on their way from Einstein's home to the Institute for Advanced Study, when Gamow casually mentioned a new idea that he had heard from another physicist, Pascual Jordan. Jordan had mentioned, tongue in cheek, that a star could be made out of nothing at all, because at the point of zero volume its negative gravitational energy precisely cancels out its positive mass energy. 'Einstein stopped in his tracks,' Gamow tells us, 'and, since we were crossing a street, several cars had to stop to avoid running us down.'

Jordan's idea will not work for the formation of a star, because any star trying to form from a singularity in this way will be inside a black hole, invisible to the Universe at large. But it *will* work for the creation of an entire universe, within the black hole. Provided the Universe is indeed closed, like the inside of a black hole, the energy required to make a universe from a singularity is indeed zero! It is, in the words of Alan Guth, 'the ultimate free lunch'. Quantum uncertainty allows bubbles of energy to appear in the vacuum, and energy is equivalent to mass. According to the rules of quantum uncertainty, the less mass–energy such a bubble has, the longer it can exist. So why should a bubble with *no* overall mass–energy not last forever?

The problem with all this, and the reason why Tryon's idea did not cause much of a stir in 1973, is that, whatever the quantum rules may allow, as soon as a universe containing as much matter as ours does start to expand away from a singularity, its enormous gravitational force (imagine the pull of gravity associated with an object containing the entire mass of the Universe in a volume smaller than an atomic nucleus) would pull it back together and make it collapse back into a new singularity in far less than the blink of an eye. What Tryon was saying, in effect, was that not just virtual particles but virtual *universes* might be popping in and out of existence in the vacuum. In 1973, he had no idea how such a virtual universe might be made real. But inflation provides a mechanism which can catch hold of a tiny, embryonic universe during that split second of its virtual existence and whoosh it up to a respectable size before gravity can do its work. Then it will take billions (or hundreds of billions) of years for gravity to slow the expansion, bring it to a halt, and make the universe contract back into a singularity.

In the 1980s, this idea of a universe being created out of nothing at all was developed by many researchers, including Tryon himself, Guth, and Alex Vilenkin of Tufts University. The consensus is that yes, indeed, universes can be born out of nothing at all as a result of quantum fluctuations. And the same powerful influence of inflation can transform any baby universe in the same way – it does not matter how much, or how little, matter goes into a black hole to make the singularity; once the new singularity starts expanding into its own set of dimensions to make a new universe, the balance between mass–energy and gravitational energy means that the new universe can be any size at all.

Bizarre though it may seem, if you could squeeze a kilogram of butter (or anything else) hard enough to make a black hole, that black hole could be the seed of a new universe as big as, or bigger than, our own. The technology is not so far-fetched, and would involve a super-powerful hydrogen bomb explosion somewhere in space at a safe distance from the Earth. It is even conceivable that our Universe was manufactured deliberately in this way, as part of a scientific experiment by a technologically advanced race in another universe.

But unless our Universe really is an artificial creation, there is still a puzzle of fine tuning, because there is no obvious reason why inflation itself should have just the right strength to 'make' a universe like our own out of a tiny quantum fluctuation of the vacuum. The 'natural' size for a universe is still down in the subatomic region where quantum effects rule, on the scale of the Planck length, 10^{-35} of a metre. And even if our Universe was deliberately designed to inflate in just the right way, where did the universe occupied by the technologically advanced race that designed our Universe come from? Somewhere, somehow, and sometime, at least one original tiny quantum seed must have grown into a universe big enough for stars and organic life-forms to exist. This is where evolution comes in.

Nobody would argue, these days, that human beings appeared out of nothing at all on the face of the Earth. We are complex creatures that could not arise by chance out of a brew of chemicals, even in some warm little pond. Simpler kinds of living organism came first, and it took hundreds of millions of years of evolution on Earth to progress from single-celled life-forms to complex organisms like ourselves.

The new understanding of cosmology suggests that something similar has happened with the Universe. It is a large and complex system, which cannot have appeared just by chance out of a random quantum fluctuation of the vacuum. Simpler universes came first, and it may have taken hundreds of millions of universal generations to progress from a Planck-length fluctuation to complex universes like our own. Lee Smolin of Syracuse University has been a leading supporter of this idea, which also takes on board notions about baby universes developed by Andrei Linde, of the Lebedev Institute in Moscow.

The key element that Smolin has introduced into the argument is the idea that every time a black hole collapses into a singularity and a new baby universe is formed, the basic laws of physics are altered slightly as spacetime itself is crushed out of existence and reshaped. The process is analogous (perhaps more than analogous) to the way mutations provide the variability among organic life-forms on which natural selection can operate. Each baby universe is, says Smolin, not a perfect replica of its parent, but a slightly mutated form.

The original, natural state of such baby universes is indeed to expand out to only about the Planck length, before collapsing once again. But if the random changes in the workings of the laws of physics – the mutations – happen to allow a little bit more inflation, a baby universe will grow a little larger. If it becomes big enough, it may separate into two, or several, different regions that each collapse to make a new singularity, thereby triggering the birth of a new universe.

Those new universes will also be slightly different from their parents. Some may lose the ability to grow much larger than the Planck length, and will fade back into the quantum foam. But some may have a little more inflation still than their parents, growing even larger, producing more black holes and giving birth to more baby universes in their turn. The number of new universes produced in each generation will be roughly proportional to the volume of the parent universe. There is even an element of competition involved, if the many baby universes are in some sense vying with one another, jostling for spacetime elbow room within superspace.

In his published papers, even Smolin has stopped short of suggesting that the Universe is alive. But heredity is an essential feature of life, and this description of the evolution of universes works only if we are dealing with living systems. I believe that the Universe – like all the universes – is literally alive. In this picture, universes pass on their characteristics to their offspring with only minor changes, just as people pass on their characteristics to their children with only minor changes.

Universes that are 'successful' are the ones that leave the most offspring. Provided the random mutations are indeed small, there will be a genuinely evolutionary process favouring larger and larger universes. Once universes start to be big enough to allow stars to form, in succeeding generations of universes there will be a natural evolution of the laws of physics. This will favour a drift in the laws and in properties like the carbon energy resonance and the mass difference between the proton and the neutron, a drift which in turn will increasingly favour the production of stars that will eventually form black holes.

Indeed, our own Universe may have been even more successful at this than I have indicated. It is possible, although by no means proven, that the cold dark matter that makes up most of the mass of the Universe may itself be in the form of a myriad of tiny black holes, each smaller than an atomic nucleus, formed under the conditions of extreme pressure that existed in the first split second after the moment of creation – or rather, the moment of birth. Any tiny irregularities *within* the superdense universe at that time may well have become sufficiently compressed to squeeze themselves out of existence, and into a new existence, as singularities inside black holes. They would show up today as rather curious subatomic particles, each with about the mass of a mountain. If they do exist, they may well be detected by the experimenters now searching for dark matter particles. Whether or not this is the case, though, there would still be continuing evolutionary pressure for the 'leftover' matter in a universe to be processed as efficiently as possible into yet more black holes.

The end-product of this process should be not one but many universes which are all about as big as it is possible to get while still being inside a black hole (as nearly flat as possible), and in which the parameters of physics are such that the formation of stars and black holes is favoured. Our universe exactly matches that description.

This explains the otherwise baffling mystery of why the Universe we live in should be 'set up' in what seems, at first sight, such an unusual way. Just as you would not expect a random collection of chemicals to organize themselves suddenly into a human being, so you would not expect a random collection of physical laws emerging from a singularity to give rise to a universe like the one we live in. Before Charles Darwin and Alfred Russel Wallace came up with the idea of evolution, many people believed that the only way to explain the existence of so unlikely an organism as a human being was by supernatural intervention; recently, the apparent unlikelihood of the Universe has led some people to suggest that the Big Bang itself may have resulted from supernatural intervention. But there is no longer any basis for invoking the supernatural. We live in a Universe which is exactly the most likely kind of universe to exist, if there are many living universes that have evolved in the same way that living things on Earth have evolved. And the fact that our Universe is 'just right' for organic life-forms like ourselves turns out to be no more than a side-effect of the fact that it is 'just right' for the production of black holes and baby universes.

Cosmologists are now having to learn to think like biologists and ecologists, and to develop their ideas not within the context of a single, unique Universe, but in the context of an evolving population of universes. Each universe starts from its own big bang, but all the universes are interconnected in complex ways by black hole 'umbilical cords', and closely related universes share the 'genetic' influence of a similar set of physical laws. The ripples in time traced out by the sensors on board COBE are, on this new picture, just a tiny part of a much more complex and elaborate structure, a structure which maintains itself far from equilibrium, and in which universes in which the laws of physics resemble those in our Universe are far more common than they ought to be if those universes had arisen by chance. The COBE discoveries do not mark the end of the science of cosmology, but the beginning of a new science of cosmology, much bigger in scope, probing further in both space and time than cosmologists could have imagined even a few years ago.

But this realization that our Universe is just one among many, that it is alive and that no supernatural influences need be invoked to explain its existence, is still not the most dramatic conclusion we can draw from this new understanding of cosmology. Although it is now clear that the Universe has not been set up for our benefit, and that the existence of organic life-forms on Earth is simply a minor side-effect of an evolutionary process involving universes, galaxies and stars which actually favours the production of black holes, nevertheless it is clear that the existence of life-forms like ourselves is an inevitable side-effect of those greater evolutionary processes.

The same laws of physics apply throughout our Universe, and throughout many other universes besides. Organic (carbon based) material occurs in profusion between the stars of a spiral galaxy like our Milky Way. This carbon-rich material may play a crucial role in the processes that allow gas clouds to cool and new stars to form. But whatever the reasons why it has evolved, it will undoubtedly seed any Earth-like planet that forms along with those new generations of stars.

Astronomers calculate that there may be as many as 10^{20} planets suitable for life-forms like ourselves in our Universe. We see the components of organic life everywhere in the Universe, and the chances are that most of those 10^{20} planets are actually alive, in the same way that Earth/Gaia is alive. The birth of the living Universe inevitably gave rise to the birth of living planets, and the seeds of our own existence are revealed in the ripples in the background radiation.

Further Reading

This book follows directly from, and develops themes discussed in, two of my recent books. They are *The Matter Myth* (with Paul Davies; Penguin, London and Touchstone, New York, 1991) and *In Search of the Edge of Time* (Black Swan, London and Harmony, New York, 1992). In an older book, *In Search of the Double Helix* (Black Swan, London and Bantam, New York, 1985) I described how evolution works here on Earth. If you want to know more about life, the Universe and everything, I recommend the following books to start with. Many of them contain further recomendations, which can lead you into an ever-widening pool of information. Some are a little heavier going than the present book, but none are dauntingly technical.

John Barrow and Frank Tipler, *The Anthropic Cosmological Principle*, Oxford University Press, 1986.

Exhaustively comprehensive, and sometimes appearing to be dauntingly technical, but full of highly readable (if you skip the equations) insights into anthropic cosmology and cosmic coincidences from the traditional perspective.

Jeremy Bernstein, *Three Degrees Above Zero*, Scribner's, New York, 1984.

A detailed account of the discovery of the cosmic microwave background radiation by Arno Penzias and Robert Wilson.

Marcus Chown, *The Afterglow of Creation*, Arrow, 1993.

The best 'instant guide' to the story of the discovery of the cosmic microwave background and the significance of the COBE discoveries in their historical context.

Francis Crick, *Life Itself*, Macdonald, London, 1982.
Crick's controversial claim that the seeds of life may have been deliberately planted on Earth; includes a concise summary of how living systems operate.

Paul Davies, *The Accidental Universe*, Cambridge University Press, 1982.
A rather technical, but brief, discussion of anthropic cosmology and cosmic coincidences from the traditional perspective.

Richard Dawkins, *The Blind Watchmaker*, Longman, London, 1986.
The best guide to the way in which evolution works.

William Day, *Genesis on Planet Earth*, second edition, Yale University Press, 1984.
A student text, but very clear and accessible to the interested non-specialist, about the origin of life on Earth.

John Gribbin, *In Search of the Big Bang*, Bantam, New York and Black Swan, London, 1986.
A pre-COBE summary of the state of knowledge about the life of the Universe.

John Gribbin, *Blinded by the Light*, Harmony, New York and Black Swan, London, 1991.
My account of how the Sun in particular and the stars in general work.

John Gribbin and Mary Gribbin, *Being Human*, Dent, 1993.
Putting people in an evolutionary perspective.

Edward Harrison, *Darkness at Night*, Harvard University Press, 1987.
The definitive history of Olbers' Paradox, including the contribution of Edgar Allan Poe.

Fred Hoyle, *Galaxies, Nuclei, and Quasars*, Heinemann, London, 1965.
An iconoclastic mid-sixties perspective on the big ideas in cosmology, including some of the coincidences discussed in Chapter Seven.

James Lovelock, *Gaia*, Gaia Books, London, 1991.
The most readable and accessible of the many introductions to the Gaia hypothesis, replete with colourful and informative illustrations. If you want to get your teeth into something a bit more meaty, try Lovelock's *The Ages of Gaia*, published by Norton in New York and Oxford University Press in London in 1988. I go into more detail about the links between Gaia and climate in my book *Hothouse Earth* (Grove, New York and Black Swan, London, 1990).

Lynn Margulis and Dorion Sagan, *Microcosmos*, Summit, New York, 1986.
Scientist mother and writer son combine to explain the evolution of life within the framework of the symbiotic theory of the origin of modern eukaryotic cells, and the importance of bacteria in maintaining life-sustaining conditions on Earth. The same team describe Kwang Jeon's work (touched on in my Chapter Five) in an article 'Bacterial Bedfellows' in *Natural History* magazine, March 1987.

Christine Sutton, *Spaceship Neutrino*, Cambridge University Press, 1992.
Neutrinos are so fascinating that I probably went into slightly more detail than I should have in my discussion of them in Chapter Nine. If you are sufficiently fascinated to want to know even more, this is the definitive historical guide, although not quite up to date as far as the cosmological implications go.

Lewis Thomas, *The Lives of a Cell*, Viking, New York, 1974.
Some of the best writing on biological themes around, including the metaphor of the atmosphere of the Earth as a cell membrane.

Lewis Thomas, *Late Night Thoughts on Listening to Mahler's Ninth Symphony*, Viking, New York, 1983.
Another excellent collection of thought-provoking essays.

Steven Weinberg, *The First Three Minutes*, Deutsch, London, 1977.
A popular account of the standard model of the hot Big Bang.

As far as I know, the only scientist who has taken the idea of the living Universe seriously enough to publish scientific papers on the subject is Lee Smolin of Syracuse University. Some of his ideas appeared in the journal *Classical and Quantum Gravity* (volume 9, pages 173–91, 1992), and in a Syracuse University 'preprint' (SU-GP-91/10–5) that had not been formally published at the time I was completing this book, late in 1992. This pair of entrancing papers, arriving on my desk in the wake of COBE's discovery of the 'ripples in time', helped to crystallize my own more fuzzy ideas about the application of Gaian reasoning to the Universe at large.

Index